Family life in old age

Character of the series

The Netherlands Interuniversity Demographic Institute (N.I.D.I.) at The Hague and the Population and Family Study Centre (C.B.G.S.) of the Ministry of Public Health and the Family at Brussels are jointly presenting this series of monographs, collections of essays, and selected articles in an effort to make the results of population studies carried out in the Low Countries accessible to international readers and research workers.
The series will not only contain studies in formal or analytical demography, but will, for example, also cover the fields of social, historical and applied demography, and that of family studies.
Manuscripts to be included in the series will be selected on the basis of their scope or methodological significance, or because they make an important contribution to demographic knowledge on the Low Countries.

Publications of the Netherlands Interuniversity Demographic Institute (N.I.D.I.) and the Population and Family Study Centre (C.B.G.S.)
Vol. 8

Family life in old age

Proceedings of the meetings of the European Social Sciences Research Committee in Dubrovnik, Yugoslavia, 19-23 October 1976, and Ystad, Sweden, 26-30 September, 1977

Edited by

G. Dooghe *and* J. Helander

Martinus Nijhoff Publishers
The Hague | Boston | London 1979

Netherlands Interuniversity Demographic Institute / Nederlands Interuniversitair Demografisch Instituut (N.I.D.I.), Prinses Beatrixlaan 428, The Hague (Voorburg), the Netherlands.

Population and Family Study Centre / Centrum voor Bevolkings- en Gezinsstudiën (C.B.G.S.). Ministry of Public Health and the Family, Manhattan Center, Kruisvaartenstraat 3, 1000 Brussels, Belgium.

The distribution of this book is handled by the following team of publishers:

for the United States and Canada

Kluwer Boston, Inc.
160 Old Derby Street
Hingham, MA 02043
USA

for all other countries

Kluwer Academic Publishers Group
Distribution Center
P.O. Box 322
3300 AH Dordrecht
The Netherlands

Library of Congress Cataloging in Publication Data CIP

International Association of Gerontology. European Social Research Committee.
Family Life in Old Age.

(Publications of the Netherlands Interuniversity Demographic Institute (N.I.D.I.) and the Population and Family Study Centre (C.B.G.S.) = v.8)
1. Gerontology--congresses. 2. Family--congresses. 3. Aged--family relationships--congresses. i. Dooghe, G., 1937- ii. Helander. Jan. 1930- iii. title. iv. series:
Nederlands Interuniversitair Demografisch Instituut. Publications of the Netherlands Interuniversity Demographic Institute (N.I.D.I.) and the Population and Family Study Centre (C.G.B.S.) = v.8.
HQ1061.153 1979 301.43'5 79-12409

ISBN 978-94-011-8385-7 ISBN 978-94-011-9080-0 (eBook)
DOI 10.1007/978-94-011-9080-0

Softcover reprint of the hardcover 1st edition 1979

EUROPEAN SOCIAL RESEARCH COMMITTEE

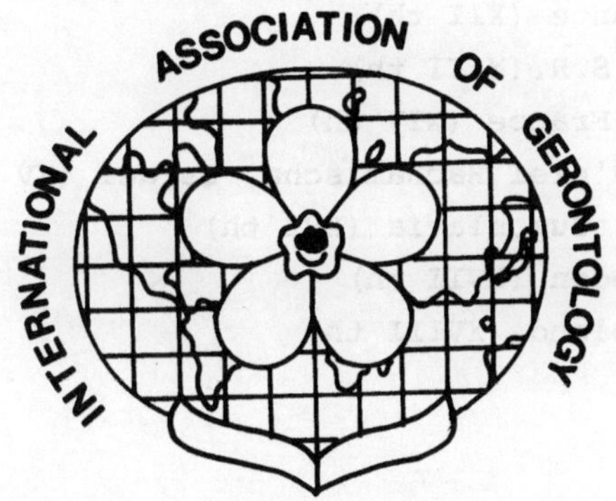

INTERNATIONAL ASSOCIATION OF GERONTOLOGY

European Social Research Committee

SYMPOSIA OF GERONTOLOGY

1954 Sheffield, United Kingdom (I st)

1956 Copenhagen, Denmark (II nd)

1957 Merano, Italy (III rd)

1959 Assisi, Italy (IV th)

1960 Berkeley, U.S.A. (V th)

1963 Markaryd, Sweden (VI th)

1966 Semmering, Austria (VII th - XI th)

1971 Paris, France (XII th)

1972 Kiev, U.S.S.R.(XIII th)

1974 Grenoble, France (XIV th)

1975 Kibbutz Ma'alei Hachamischa, Israel (XV th)

1976 Dubrovnik, Yugoslavia (XVI th)

1977 Ystad, Sweden (XVII th)

1978 Krakow, Poland (XVIII th)

PUBLICATIONS

1. 1956 *Copenhagen*
The need for cross-national surveys of old age. Report of a Conference at Copenhagen, October 19-23, 1956. European Section Social Science Research Committee, International Association of Gerontology. Published by: Division of Gerontology, University of Michigan, An Arbor, Michigan.
2. 1957 *Merano*
Aging and social health in the United States and Europe. Report of an International Seminar held at Merano, Italy, July 9-13, 1957. Published by: Division of Gerontology, University of Michigan, An Arbor, Michigan.
3. 1966 *Semmering*
Vol. 1: Sheila S. Chown and K.F. Riegel(ed.): Psychological Functioning in the Normal Aging and Senile Aged. 1968, Basel-New York.
Vol. 2: J. Madge and E. Shanas(ed.): Methodological Problems in Cross-National Studies in Aging. 1968, Basel-New York.
Vol. 3: Marjorie F. Lowenthal and A. Zilli(ed.): Colloquium on Health and Aging of the Population. 1969, Basel- New York.
Vol. 4.: A.T. Welford and J. Birren(ed.): Decision Making and Age, 1969, Basel-New York.
Vol. 5: L. Gitman and E. Woodford-Williams(ed.): Research Training and Practice in Clinical Medicine of Aging 1970, Basel-New York.
4. 1971 *Paris*
Elderly people living in Europe. Report of the European Social Research Committee (of the IAG). Published by the International Center of Social Gerontology, Paris 1972.
5. 1975 *Grenoble* / 1976 *Jerusalem*
Dependency or interdependency in old age, The Hague 1976.

Contents

Preface

The Netherlands Interuniversity Demographic Institute (N.I.D.I.) and the Population and Family Study Centre (C.B.G.S.) evidence the growing importance attached to the field of social gerontology. The two institutions are designed to coordinate and to stimulate all kind of research in the field of population and family.

Long-run trends in demographic processes of mortality and fertility have had consequences for the kin network. The increasing number of aged people in the total population and the reduced number of descendants to whom an older person may turn for assistance is becoming a real problem in Western society. The problem of the Elderly is too important in order to be neglected.

Volume VIII of the N.I.D.I.-C.B.G.S. publications contains a number of articles concerning the family life in Old Age. The European Social Research Committee on Ageing held two colloquia on this topic. The papers presented at the Dubrovnik meeting, Yugoslavia 1976, and at the Ystad meeting, Sweden 1977, are published in this volume.

The editors hope that this volume, the eighth in their yearly publication series, will serve to give more insight in the complex problem of the elderly in our society and hope that more cross cultural research will be undertaken.

The editors

Introduction

After the monograph entitled 'Dependency or interdependency in old age' (Munnichs, Van den Heuvel, 1976) the European Social Sciences Research Committee (ESRC) of the International Association of Gerontology has prepared the present book Family Life in Old Age. This volume is the product of symposia held in Dubrovnik, Yugoslavia, in October of 1976 and in Ystad, Sweden, in September of 1977.

The main function of the ESRC is to co-ordinate relevant research work in the field of gerontology. The ESRC has always tried to stress the importance of the multidisciplinary approach. At the symposia, in which the authors of the various chapters of this book participated, research results were presented, criticized, and revised. The symposia of the ESRC should be seen as an experiment in cooperation and integration of all kinds of researchers and research in the field of social gerontology. The participants have regularly the opportunity to meet and present papers on a theme chosen by the members of the Committee.

The ESRC symposia in Dubrovnik and Ystad were focussed on the theme of Family Life in Old Age. Systematic information about the many aspects of this problem is lacking, and much more research is needed. The subject matter of the two symposia covers an area that is of great importance for research on social gerontology. Although the topic is exceedingly broad and the participants represented a wide variety of disciplines, this book is characterized by both diversity and unity. The different viewpoints of epidemiology, the social and psychological sciences, and clinical research are centred on the topic of family life in old age.

It was also the intention of the members of the ESRC that the results of the symposia will be published regularly. The ESRC wish to express their gratitude and appreciation to the Population and Family Study Centre (C.B.G.S.) in Brussels and the Netherlands Interuniversity Demographic Institute (N.I.D.I.) in The Hague for making it possible to publish the papers of the symposia in the N.I.D.I.-C.B.G.S. publications. Responsibility for the contents of the papers lies with the authors.

For convenience, the selected contributions are grouped in four sections: general topics, relations inside the family, relations outside the family, research and methods.

GENERAL TOPICS

In her communication entitled *What is a family?*, E. Melin (Sweden) discusses a number of conceptual problems. Cross-cultural comparisons give rise to problems when such concepts as family, relatives, and household are not clearly defined. How are they to be circumscribed despite the overlapping?

P. Paillat (France), in his paper *Influence of demographic trends upon family building and elderly people's role*, underlines the changes in intergenerational relations resulting from demographic changes. The longer average lifespan has given rise to a new problem, i.e., the presence of two elderly generations in the same family. The concept life-cycle also changes continuously. At present, the phase of grandparenthood lasts on average 15 years and thus introduces a new phase into the family cycle. Economic, social, and political changes also exert an influence. The concepts family life cycle and nuclear family are examined critically in the paper on *Ageing and sociological studies of the family* by W.R. Bytheway (Great Britain). A retrospective longitudinal study done in 934 families has shown that the family life cycle, as defined in family sociology, is not representative. Only 22 per cent of the families have an 'empty-nest' period. The author makes a strong case for more longitudinal research.

RELATIONS INSIDE THE FAMILY

H. Weihl (Israel) puts forward a number of considerations concerning the relationship pattern of the elderly and their children. According to her, the study of these interrelationships should be based on an historical perspective. H. Worach-Kardas (Poland) reports in her contribution called *Family and neighbourly Relations - their role for the elderly* that in Poland many parents and married children still live together, but in recent years this phenomenon has been undergoing a rapid change. The oldest generation are contributing to this change too, because of a reluctance to become over-dependent on the children. The papers of A. Ciuca (Rumania), *The elderly and the family*, and A. Tymowsky (Poland), *The influence of old people in a family on its living standard*, also underline the leading role of the elderly in their family. The Rumanian article indicates, furthermore, that active participation in family life is directly correlated with the feeling of satisfaction of the elderly. Tymowski points out that the time budget of the aged has not yet been sufficiently discussed with respect to the services rendered to and by the family. More research on the evolution of the consumption model of the aged; especially as they become older, is necessary. The contribution of H. Olsen (Denmark), *Family contacts and social class in the early stages of old age*, is based on the findings of a longitudinal study (9 years, 4 surveys) and provides some insight into the period before and after retirement. The results show, for instance, that during the study period a change occurred in the number of family contacts according to social class, the lower classes ending with fewer contacts than the higher classes.

Furthermore, gradual retirement, which occurs more frequently among the higher classes, has a more positive effect on family contacts and any degree of isolation than does the much more sudden retirement usually occuring in the lower classes. On the basis of an attitude study in 100 middle-aged women, Schlettwein-Gsell and Bass (Switzerland) stress, in their paper on *Social needs of the relatives of old people,* the need for more information about services for the aged.

RELATIONS OUTSIDE THE FAMILY

Several of the papers are concerned with the problem of the isolation of the elderly. Under the title *With or without a family,* M. Asiel (Belgium) discussed some of the results of a study performed in Brussels on the living conditions of women aged 70 and older and living alone, with special attention to familial and other social relations, particularly with respect to such problems as loneliness and isolation. Special attention is also given to loneliness in the communication of G. Dooghe and L. Vanderleyden (Belgium). The paper entitled *Loneliness of old widows and married women* concerns the use of a path-analysis model to determine the effect of eight predictors on the isolation of these groups. The highest predictive value was found for the subjective health evaluation and being alone frequently or infrequently.

The contribution of K. Knipscheer (The Netherlands), *The primary relations in old age,* deals with a comparative analysis of five categories of relationships maintained by the aged, i.e., with children, with brothers and sisters, with other relatives, with friends, and with neighbours. Besides the parent-child relationship, the relationships with friends and neighbours proved to be of great importance, particularly with respect to affection, assistance, and free time. In relation to the progressive ageing of the population, M. Dieck (West Germany) discusses, in her paper entitled *Typology of the need for community services under the aspect of civil status and family relations of the elderly,* the need for various provisions for the elderly in the light of the increasing desire of both the aged and the government that this group remain in their own homes living independently as long as possible. In the paper on *Family helping patterns in a social Swedish retirement club,* L. Tornstam points to the strong negative relationship between the help parents ask their children for and the number of friends they have, which may indicate that in a number of cases the contacts with friends form an alternative for help from children. L. Dahl (Sweden) deals, in her paper on *The geriatric Ward and the Family,* with the communication between the personnel in a home for geriatric care and the patients' family. The quality of the interaction between members of the family, the personnel, and the aged patient determines to an appreciable degree the rehabilitation of the patient. The main question concerns how the intensity of this interaction can be improved.

RESEARCH AND METHODS

That personal development is still possible at advanced ages is shown by the communication by N. Stevens and M. Wimmers (The Netherlands) on the basis of experiments with *encounter groups with elderly persons.* The initiative for the formation of such groups should become more general, and might be assumed by service centres. The training of group leaders remains a problem, and the author wonders whether former members of encounter groups could be trained so that they in turn could assume the role of social worker. In this connection reference is made to the Continuum Center for Adult Counseling and Leadership Training of Oakland University in Michigan.
A.M. Bevers (The Netherlands) describes a research program on *The awareness-context of the intergenerational relation.* To obtain more insight into the relationship between aged parents and their children, it seems important to know what the parties involved know about each other. Informa-

tion about the entire combination of what each interactant knows in a given situation about the identity of the other and about his own identity in the other's eyes is required for recognition of the qualitative aspects of relationship.

Agreement has not been reached on the question of whether there is solidarity between the generations. To start with some of the positive points, a larger number of families belong to the three- and four-generation family category, and the change of having contacts with grandchildren has increased compared to the period when life-expectancy was limited. Marrying at younger ages, in combination with a smaller family size, can lead to less separation between the ages of the generations. On the other hand, increased mobility - particularly from rural areas to cities - seems to have a negative effect, as does the acceleration in the genesis of new ideas, which increases the generation gap. Thus, some situations indicate intimacy between the generations, albeit at a certain distance, whereas others indicate that the gap is increasing. Consequently, many of the papers argue for more closely focussed longitudinal research to supplement and improve the available knowledge in the field of gerontology.

G. DOOGHE

J. HELANDER
Chairman

General topics

1. What is a family?

Else Melin (*)

Intrigued by the title of this symposium, I have attempted to arrive at a definition of "family". The task, undoubtedly proved to be a difficult one in view of the existing cultural differences which predominate in different parts of the world. People often speak of the "Western culture", the "European culture" and so on. What precisely does "culture" mean? In his book "The Children of Sanchez" Lewis (1961) defines culture as habits of living that are transferred from generation to generation. Foster (1965) says it is an integrated, functional unit. Whereas Clark (1959) points out that people are members of a group of relatives, to whom they are responsible for their behaviour and on whom they are dependent for support and social sanction. The differences between the behaviour in different tribes has been shown very well by Mead (1939). Behaviour is something we learn. It is a form of communication.

The social factor, too, is very important. We know that personal identification with small groups seems to be necessary for most people since it provides psychological security and satisfaction in their daily work. People need to be with others. In every society people learn the behaviour that is appropriate to them and which they expect from others, in the infinite number of situations in which they find themselves (Foster, 1965). However,

(*) Gerontologiskt Centrum, Lund, Sweden.

cultures change, more or less rapidly. When cultures change, the norms and values as well as the behaviour change. Very often old people are slower to change than the young; as a result, the old are looked upon as obsolete or old-fashioned.

The cultural changes influence the individual's role perception. These culturally-determined behavioral expectations and the differential expectation of roles, can cause many problems between generations (Mead, 1970). One example is communication problems. Communication means that someone initiates actions in the form of symbols: verbal, visual or both combined. Someone else interprets these symbols in accordance with a culturally determined understanding of what they mean. One of the basic functions of culture is to facilitate communication (Foster, 1965). When people are exposed to the symbols of a culture other than their own, these symbols are often misunderstood or not understood at all. Motor patterns and customary body positions are examples of aspects of culture that people learn. It is easy to misinterpret motor patterns in a culture different from one's own.

Young people can easily change their behaviour when they meet influences from other cultures. (Sometimes we also speak of a special "youth culture"). Young people may also have been at school much longer than their parents and other relatives which may cause communicational problems because of educational differences. Words and symbols also acquire different meaning from time to time and the language differs between generations. New words and new ideas mean that people must learn and "unlearn". Since it takes more time for the older to unlearn and learn, problems can arise (Toffler, 1970; Mead, 1970).

Can we speak of sub-cultures in Europe? Are there sub-cultures in a country? Do different generations represent different sub-cultures? Do people maybe change in ways that suit their interests?

Helander (1973) has pointed out that it is much easier to learn if you have background-experience in the field you are pursuing. He compares different persons and their possibilities and abilities by means of a figure.

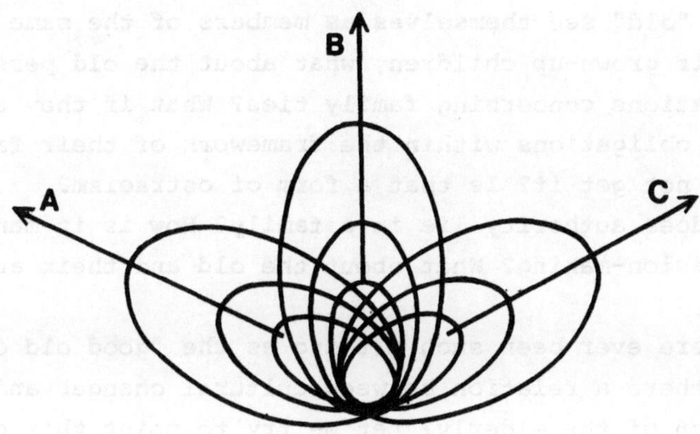

This figure has three parts: A, B and C. Each part represents a person. The ovals grow bigger and bigger for each of them from youth to old age. The experience is growing in a head-direction for each person. The broadly humane is what all persons have experience of and that is the area which is common for A, B and C. This area may grow with age. There are also areas which are individual and not shared with anyone else. These areas may also grow with age.

Do different generations perhaps mean different things when they use the word "family"? Let us consider some questions which might throw some light on this subject:
What is a "family"?
Who belongs to it?
Is it possible to leave a family?
Is it possible to change family-membership?
Can *one* person be a family?

Can people live without being a member of a family?
How many families can you be a member of?
Who decides whether a person belongs to a certain family?
Can you belong to a family though you do not want to?
Can a family see a person as member of it, though he has left it?
Can old people have other opinions than young people concerning family membership?
If **the "old" see themselves as members of the same family as their grown-up children, what about the old person's expectations concerning family ties? What if they expect mutual obligations within the framework of their family and do not get it? Is that a form of ostracism?**
Where does authority lie in a family? How is it manifest in decision-making? What about the old and their authority?
Has there ever been such a thing as the "good old days"?

Is there a relation between cultural changes and the position of the elderly? Let me try to point this out on a continuum.

Urbanization process ⟶

Rural culture	*Urban culture*
"Old are wise"	"Old are incompetent"
The old have authority	The middle-aged have authority
Few old persons	Many old persons
Family responsible for nursing the old	Society responsible for nursing the old

Is it possible to define some steps on this continuum?
Where is your country on the continuum?
What criteria can we use to define the position for a culture on the line?
What are the relations between the urbanization process and the social structure in family life?
When does a family see the old person as incompetent? Which criteria are used?

Is there some crisis which it is possible to prevent?
Could there be new roles for the pensioneers in the urban culture?
What are the responsibilities in the family in different cultures?

These questions about "family life" may evoke different answers. Yet, while the construct "family" has different meanings in different sub-cultures in Europe, those of us present at this meeting have the advantage of being "European". Our common meeting ground, the English language, helps somewhat to eliminate any severe communication problems. But, since we are inclined to put different meanings in the same construct, I would suggest that we try:

1. To make a *construct validation* of the construct "family".
2. To find criteria to identify the sub-cultures to which our countries belong.
3. To describe where our countries are in the urbanization process, especially concerning the old people's situation.
4. To find mutual concepts so that we could *really* understand each other.

I would like to propose a few other ideas for consideration. Since the Latin word "familia" means "all the servants of the house", what does the Latin expression "pater familias" mean? Does it not connote "father of the servants" or "the serving father" or does it perhaps have still another meaning?

Likewise, the Greek word "oichos" meaning "house" is used in words like "ecology" and "economy". Since-logy and -nomy mean "the science of", it would follow that "ecology" and "economy" refer to "the science of the house".

By definition "ecology" is the science of man and environment at interplay with each other. Does that imply that the "family" (= the servants of the house) shall serve *nature* and *environment*? These considerations, undoubtedly, call for reflection upon the theme selected in this paper: "What is a Family"?

ABSTRACT

This paper is an attempt to find a definition of "family". The author has posed many questions, but does not answer them. Suggestions are made that the members of the symposium try to make a construct validation of the construct "family" to try to find criteria to identify the sub-cultures to which our countries belong, to try to describe where our countries are in the urbanization process and to try to find mutual concepts in order to understand each other.

REFERENCES

Clark, M., *Health in the Mexican-American Culture*, Berkeley and Los Angeles University of California Press, 1959.

Foster, G.M., *Traditional Cultures and the Impact of Technological Change*, Harper International, New York, 1965.

Helander, J. "Åldrandet - åldrandena" in Melander, S.: *Äldre idag - i morgon*, Verbum, Stockholm, 1973.

Lewis, O., *The Children of Sanchez*, 1961.

Mead, M., *Sex and Temperament*, From the South Seas, Morrow, 1939.

Mead, M., *Male and Female*, New York, 1949.

Mead, M., *Culture and Commitment: A Study of the Generation Gap*, New York, Doubleday, Natural History Press, 1970.

2. Influence of demographic trends upon family building and elderly people's role

Paul Paillat (*)

It is obvious that the dramatic decline of death rates bears directly upon family-building (understood as both creation and composition); the decline of birth rates is also relevant, even if it is less obvious. We need to understand how both phenomena act since they are not independent. In this paper we will try to demonstrate how the role of elderly changes when their own family passes through deep modifications as is the case in European populations. Taking into consideration a smaller death rate decline, a higher level of fertility, a variety of family structures (including polygamy and consensual unions) would make the analysis much more complicated. However we should not overlook the fact that in developping countries the status of elderly people is declining, especially in urban areas.

Let us see first what the concept "family" implies. It has two main meanings. In the first one, the wider and the less precise, *a family at a given time includes every individual related, either by consanguinity, (kinship, lineage) or by marriage (marriage between two representatives of different lineages).* For our purpose, we need to limit our observation field to the patrilinear structure as do genealogists: it is indeed the proper way for computing the distance to a common ancestor. To follow a matrilinear structure would not modify the analysis to come but would be less consistent with the Western usage. Generations cannot be replaced without the

(*) L'Institut National d'Etudes Démographiques (I.N.E.D.), France.

contribution from external people (sons- and daughters-in-law) paying attention to rules governing incest; this replacement takes place according to a conducting thread: in this meaning, we use to talk about three- or four-generation families (i.e. co-existing, if not co-living) or *kin-family*. In most cases, however, we will pay attention to *nuclear family*, constituted by a couple and its unmarried and co-living children. French statisticians call it also *biological family* but one may wonder why: how can a lineage be non-biological ? "Biological" has only one meaning here as opposed to "adoptive". When necessary we will explicitly mention either *kin-family* or *nuclear-family*.

KIN-FAMILY IN HIGH DEATH RATE CONDITIONS

When expectation of life at birth was around 35 or 40 years, as was the case in European populations during the XVIIIth century and even the XIXth and as it was also in developing countries in the middle of the XXth century, the three-generation kin-families were rare. To reach that step, one needs to combine the probability to be 45-year-old (for a man) with that of having a surviving son, himself father of a boy. Of course a high fertility increases the probability of having both son and grand-son but their co-existence has no practical meaning if it does not last some time. As population historians indicated, patriarchal families were largely mythical: their rarity pushed them up to the rank of models. Is a common situation a model ? For the same reason a very old man cannot be but respected by all: he was under divine protection since he has not been submitted to common fate.

As a complementary observation let us mention that in France during the XVIIIth century, one was orphan on both sides at the average age of 29.5 and not at 55 as to-day.

KIN-FAMILY IN LOW DEATH RATE CONDITIONS

Nowadays, male expectation of life at birth reaches 70

years, a level which increases the probability of being a grand-father. With an age at marriage of, say, 25 years, one has a fair chance to have a grand-son (or a grand-daughter) at 52 or 53 but 25, 26 years more are needed in order to be a grand-grand-father and less than 4.200 men reach that age out of 10.000 newly-born males (French life table, 1973). On the family side, two factors play a role: on the one hand, a longer life-span and, on the other hand, a lower average age at marriage. It is then common to be a grand-mother at 42 or 43 (where are the white-haired females traditionally associated with this statute ?), and a grand-grand-mother at 65. This new family calendar of events implies co-existence of two generations of retired people: a 80-year-old widow with a 60-year-old daughter, herself a widow, will be a less and less exceptional case. How can a 38-year-old grand-daughter, with depending children, be able to help her mother and grand-mother ? Without another lowering of age at marriage, it is possible to observe five co-existing female generations, even if it is still unusual.

NUCLEAR FAMILY IN LOW DEATH RATE CONDITIONS

Let us observe now the case of a nuclear family. The use of such a term is meaningful: in order that the term "family" covers often a couple with its children under age (or at least unmarried) a deep change in morals is necessary, including the end of co-living arrangements which still prevail only in farm-rural people. As mentioned before, co-existence is possible without co-living: in other words a couple in which the husband is a father (i.e. two generations) at, say, 30 years and the wife is 26-27-year-old, has normally also its four parents (i.e. three generations); ten years later, one of the grand-fathers is likely to be dead; twenty years later, the first generation will be only represented by the grand-mothers (some of them being then grand-grand-mothers).

Using another approach, a French demographer, Le Bras (1973) confirms in some way what has been said. He resorted to simulations using parameters corresponding to France in 1972 and in the XVIIIth century. In the framework of appropriate stable populations he reached the following results: at birth, 89 per cent of children have both grand-mothers; 42 per cent have their four grand-parents; at 21 years, 17 per cent have lost every grand-parent but have still their own parents; other cases are distributed among 63 possible combinations. In the XVIIIth century, 12 per cent of 21-year-old-children had their father, their mother and a grand-mother. We will refer again to figures or observations from Le Bras.

Such a change of the kin-family chronological history modifies affective loads and often material burdens in the nuclear family. The modification may be a positive one when, for example, grand-mothers help their daughter and daughter-in-law; it may be on the negative side when the daughter or daughter-in-law bears too heavy a burden between her own home care and frequent visits to the mother or mother-in-law (or both), living nearby but in poor health conditions. From surveys even in urban areas, we know that a significant percentage of old people live in the vicinity of their adult children, if not under the same roof, but we will come back to this point later on.

NUCLEAR FAMILY IN LOW FERTILITY CONDITIONS

Up to now we have not paid attention to the fertility trend in spite of the fact that to-day an increased life-span is leading to the gates of very old ages. More and more people are getting along with a sensible decrease of off-spring which leads to the non-replacement level, unit by unit, of successive generations. In other words, when lower death rates bring to a large majority of men and women the possibility of beeing grand-parents, lower birth rates restrict seriously such a prospect.

What are the conditions to be met in order to be a grand-father or a grand-mother ? All depends upon a probability chain: first, the probability of surviving until the age when one may be a grand-father or a grand-mother, taking into consideration the modal age at marriage; then the probability of having a child (or at least one of the children) who becomes father or mother. When we compare two 55-year-old men of whom one is the father of a single child, and the other the father of two, the latter has a double probability of eventually being a grand-father. In other words, when the second generation is with the first one in a ratio equal to or a little over one, the probability of reaching such a state is reduced in due proportion. When the third generation is also linked with the second one by an even lower fertility rate, again the mentioned probability will be smaller. Of course, historical events act upon this kind of probability from one period to another. For instance, in France, marriage cohorst of the 1950's, and 1960's had, on average, substantially more children than promotions 30 years older (those of their parents), and curiously enough, these Malthusian promotions were more frequent at the top of a three-generation family, than would have been the case if their children had had a similar behaviour regarding procreation or, conversely, the 1950 marriage cohorts are likely to be less often grand-parents since their own children have a more limited descent.

DURATION OF GRAND-FATHERHOOD

Anyhow, what is the matter with being a grand-father at a given time which may be very short (case of grand-son who dies a little after being registered as born and fit to live). What is important is playing this role a few years. Then the more significative concept is the length of co-existence and not an instantaneous one. Again we introduce here death rates and fertility rates to which we should add nuptiality patterns (mean or modal age at marriage). Our goal is different: how *long* will someone be a grand-father and not *who* will be one ?

Relying upon Le Bras figures, a "child" (the third generation), 20-year-old, has still two grand-parents, at least, who, reciprocally, held this role during twenty years (French data, 1972), but we should say "grand-mothers" because it is the most frequent case. From 10,000 55-year-old men, only 5,041 reach the age of 75 against 7,289 women of the same generation. Grand-fathers are taken from this group when they live from 10 to 20 years in this family status. A man who lives 25 years as a grand-father has a fair chance of becoming grand-grand-father. But, let us be frank, a woman has a much bigger chance to do the same since 30-year-old children (4th generation) have still a grand-grand-mother, at least. In other words, most of the grand-parents die when their grand-children are teen-agers or at the beginning of their active live (Le Bras).

CONSEQUENCES UPON RELATIONSHIPS BETWEEN GENERATIONS

As a first example of consequences originated by the new life calendar, we may mention with Le Bras, that the under-taker, in the Schumpeterian meaning, i.e. the one who risks his own capital into an undertaking, is likely to be more often a middle-aged or even an aged man, and that innova-tion, the spring of economic progress, is probably rarer or weaker. Is it the key explaining the success met by non-owning directors (the so-called "managers")?

In the same line of thought, the longer the grand-parents live, the later comes the heritage. Where a son is to be the man in charge of an entreprise after his father, he has to wait 40 years (median age): current migration from farms is no more a response to demographic pressure but may be partly due to this change in life calendar. Furthermore, when a patrimony is getting a new owner, it jumps in some way over a generation since the heirs have already reached such an age that this capital will of more benifit to their children than to themselves.

Does a longer duration of grand-fatherhood have an im-pact upon family daily life ? Yes, provided residences are close. According to survey figures, a high proportion of parents (first generation) live in the same "commune" than their children and grand-children: 30 per cent of parents aged 50 to 80 in a survey conducted by Roussel (1976); 32 per cent of parents aged 65 or more in non-farm rural populations, as shown by a survey made by Maslowski-Paillat (1973). European studies provide similar figures.

Neighbourhood does not mean co-living; neither does it mean separation. Roussel (1976, 240) states that "household autonomy is not related with family breakdown, but solidarity has no more the same forms, nor the same meanings".

As a concluding remark, we would agree with Roussel (1976, 249) that "the family in a wide meaning is made up with households weaker than before, that children are fewer and fewer and, above all, that relationships between generations are partly explained by a lessening of social relationships". After this comment, let us not overlook the fact that bureaucratization is the only substitute offered by our societies when the kin-family is broken up (missing links) or distorted (migrations).

SUMMARY

The influence of mortality upon family-building is discussed for three cases: kin-family and high death rates, kin-family and low death rates, nuclear family and low death rates. Then comes the role of low fertility in a nuclear family.

The probability (and its evolution) of being a grandfather or a grand-mother is the matter of the second part of this paper which ends by comments upon the above developments and the relationships between generations. At a time when it is more likely to be not only a grand-mother but also a grand-grand-mother during a longer period of time these relationships depend to a greater extent on the distance between residences of relatives, a sociological fact. Demographic factors are not the only determinants but should not be overlooked when facing such deep changes in family patterns.

REFERENCES

Le Bras, H., Parents, grands-parents et bisaïeux, *Population*, 1973, 1, 9-38.

Maslowski, J., and P. Paillat, *III. Les ruraux âgés non agricoles*, INED/PUF, Paris, 1973,(INED, cahier no. 68).

Roussel, L., *La famille après le mariage des enfants. Les relations entre générations*, INED/PUF, Paris, 1976, (INED, cahier no. 78).

3. Ageing and sociological studies of the family

Bill Bytheway (*)

One can only go so far in understanding life in old age in ignorance of the life that has preceded it. For this reason the study of any aspect of old age eventually has to give serious consideration to the historical and biographical circumstances that lead people into the situations in which they are found. It follows that any conference or book dealing with family life in old age has to consider family life in general, and more particularly the family life that has characterised the lives of people adjudged to be old: their changing family circumstances as they have progressed from childhood into adulthood towards the termination of their lives.

These kinds of concerns were expressed most forceably in Johnson (1977), a paper delivered at the 1975 meeting in Jerusalem of the European Social Science Committee of the International Association of Gerontology. In proposing a biographical approach to the study of ageing, he is reflecting a wish which many share to redirect empirical research away from static and generalised theories concerning the progression of individuals through the ageing process, towards a theoretical framework which is more relevant to the analysis of empirical records of the lives of actual people. A classic example of the former theories is the family life cycle. In the paper delivered at the Ystad Colloquium, I

(*) Medical Sociology Research Centre, Swansea.

included a critique of many features of the standard conceptualisations. I introduced this with the following statement: "The circumstances of the prematurely deceased, the childless couple, the one-parent family, the unmarried parent, the middle-aged never-married (homosexual and heterosexual), the divorced, the young widowed, the step-child, the orphan, the married couple of different ages, and the old parents of the new-born infant (all well-known interest items, let it be noted, in the popular press), should not be conceived to be pathological deviations from normal family development. *Such circumstances are the consequences of normal patterns of family development.* Given normal mortality rates and normal marital and sexual patterns of behaviour, it is *inevitable* that some people will find themselves in such circumstances.

"Given that the standard family life cycle is not only the family sociologist's model but also the desired events for most members of 'ordinary' families, the occurrence of such circumstances is widely recognised to be critical to family histories. Because of the accumulation of experience which comes with age and the ascription of success to survival and the production of descendents, such concern will be particularly characteristic of the older members of the family. Consequently the study of the family - and in particular the study of it in respect to old age - has to define the family in such a way that such deviant characteristics never remove families from the field of study".

In the review of classical family sociology that followed, I tried to demonstrate the inconsistencies and inadequacies of many definitions of the nuclear family and of the cycle in regard to the deviant statuses listed. It should be noted that all these statuses are defined exclusively upon the following phenomena: age, marriage, birth and death.

This paper is based upon the second part of the paper presented at Ystad, and it simply reports an analysis of some statistical and demographic data, drawn from a rich source of information on the histories of actual families: "Burke's Peerage". This magnificant document records the dates of birth, marriage and death for most families descended from eighteenth century British peers up to the present day. Nearly half a million persons are listed. The dates moreover include the day, month and year which enables the researcher to establish the order of events occurring within the same year and to accurately measure the interval between successive events. Before proceding to the analysis, however, I wish to stress that my primary concern is to empirically document the sequences of events that actual families experienced. From this and from a desire to see further research being focussed upon the biographies, actual, perceived or claimed, of real families and upon the effect of biographies on current and future family processes, I proceed to offer an alternative frame of reference to the family life cycle.

The sample I have analysed contains 934 families, each characterised by a married couple and the children (if any) who are listed as having been born of the marriage. In other words, I have been happy to take advantage of Burke's method of documenting the membership of families. I do however have certain reservations about its completeness. Infants who only lived for a day are frequently recorded, but I am doubtful that all such brief lives are included. Similarly one or two children (whom I in turn included in my data) are described as adopted, but I suspect that this information is volunteered by Burke's informants rather than actively sought by the editors. The important point about accepting the data as it stands is that all persons listed are clearly and unambiguously ascribed to a specific marriage, and as such are being presented by the informants to readers of the book (who will include other informants, relatives, etc.) as being members of the family that is defined by that marriage.

The 934 are all *completed* families in the sense that *either* the last surviving parent died between 1955 and 1970 (the date of compilation) *or*, in the case of marriages that were terminated by divorce, the husband died between 1955 and 1970. The reason why this identifying mechanism is different according to whether or not the marriage ended in divorce is that Burke omits the date of death of women who marry into the Peerage and who are subsequently divorced. The husband/father is the person through whom kinship links to the Peerage are obtained and so the same problem does not occur in respect to men. The 934 are *representative* in that the number of children and the length of time between any two events (e.g. the parent's marriage and the death of the last surviving parent) in no way affects the chances of a family being included. This methodological issue is discussed in the context of this data source in more detail in Bytheway (1974). The significant methodological point is that the family is identified by a single event, an event which each family experiences only once and which can be construed to be a terminating event. Thus it can be described as a "retrospective longitudinal sample", having none of the problems of bias which characterises samples of families identified through cross-sectional surveys (Bytheway, 1977).

The median date of marriage of the parents is 1910, and 90 % of all the 934 parent marriages occurred between 1890 and 1935. Thus the children - the surviving members of these families - are typically in their fifties and sixties in 1977. Their own families will all have begun to disintegrate as their children have married and as they themselves have died or become widowed. The data, however, refers to their families of origin - the surviving contemporaries of their parents are now in their eighties and nineties.

The procedures for selection, designed to ensure 'structural' representativeness (as indicated above), also meant that no British Peers were included either as parents or as children. The 934 husband/fathers were all younger brothers or more distant relations of Peers (nephews, cousins, etc.) and they never succeeded to a title of their own. Thus the sample is not aristocratic in the narrow sense of the word. Nevertheless it is undeniably drawn from the highest strata of British society. The few working class Life Peers resulting from recent nominations by Labour Governments, although included in Burke's Peerage, have not been able to supply the editors with detailed family histories. This strong social bias in respect to the total British population can in no way be belittled. It is probable that the 934 families will each have enjoyed greater chances of longevity, more effective family planning and greater opportunity to bring about a divorce than the vast majority of the total population. It has been argued that these biases mean that such cohorts in respect to demographic characteristics are simply 'ahead of their time': that their life expectancy, family planning and divorce rates are similar to the same features of the present day British population (Hollingsworth 1964). However, the objective of this paper is not to attempt to explain the observed data, but rather to consider the observed variability within the sample and to review the notions of the life of typical families in the light of the presented data.

There is one further feature of these 934 histories which cannot be ignored: most of the 934 straddled the 1914-18 and 1939-45 wars. The editors of Burke's Peerage appear to have systematically distinguished all deaths resulting from military service and as a result it can be reported that 35 of the 934 husband/fathers died in this way; 27 in the earlier war and 8 in the second war. Similarly 90 of the sons, 14 and 76 respectively, died through military action in the two wars. The question of whether or not this depletion 'invalidated' the whole sample was given serious consideration. It certainly frustrates one's inquiry in that the objective is to study family processes and such premature deaths are 'irrelevant' and 'unnecessary' interruptions of such processes. This undoubtedly will be how such deaths will have been and will still be seen by those who were bereaved. The fact is however that family processes are 'interrupted' in a wide variety of more subtle ways, each of which is specific to historically based cohorts. Methods of family planning, opportunities to terminate pregnancies, divorce laws, the social constraints upon courtship and developments in medicine are such areas in which there have been considerable changes over the last hundred years affecting the process, or the historical realisation, of individual families. The major difference being that the effects of such factors upon the individual family is less apparent than in the case of deaths in the course of military service, and are never the subject of organised public grief or celebration. On these grounds it could be argued that such historical effects including the consequences of wars are not interruptions of 'normal' processes but are rather an integral component. In order to analyse variability in family histories it is simply necessary to obtain one reasonably large, broad-based and representative sample of completed families. The fact that the sample is drawn from a certain narrow strata of British society and is subject to the various consequences of the events of a relatively brief historical period limits *only* the extent to which the observed characteristics of the sample can be extrapolated to other populations.

DATA

Within the 934, a total of 264 produced no children (28.3 per cent). Of these childless marriages, 49 (18.6 per cent) ended in divorce. Of the 670 marriages which produced children, a further 67 (10.0 per cent) also ended in divorce. Unfortunately Burke does not usually record the subsequent marital career and death of a divorced woman who had married into the Peerage, it only dates a divorce by at best the year and not the day as in the case of births, marriages and deaths, and it does not record whose responsibility the children of a divorced marriage subsequently became. For these reasons it is not possible to carry out a detailed analysis of the part that divorce has played in these family histories. In only four of the 67 divorces involving children, had a child already left the family through marriage. Thus in regards to this sample it is appropriate to class divorce with childbirth as an event predominantly associated with the early years of marriage, even though it is a permanently available means of terminating marriage.

Childlessness is a different phenomenon in that all marriages in this sample are initially childless - according to the record. A marriage only remains childless if it is interrupted by divorce or premature death, or if children are, for whatever reason, unforthcoming. Within the 215 childless families which were terminated by death, only 37 lasted for less than ten years. It is possible that a number of divorces resulted from marriages in which children would not have been produced - indeed signs of childlessness may have been instrumental in leading to the divorce - and so it is fair to conclude that something over 20 per cent of the 934 marriages did not or would not have produced children given a sufficiently long life (i.e. period of exposure).

Of the 603 families which produced at least one child and which avoided divorce there were 24 in which the data was incomplete. In the remaining 579, the precise sequence was obtained of all first marriages of family members (first remarriages in the case of widowed parents) and all deaths up to the death of the surviving widowed parent. Infants who died before the age of five years are ignored on the grounds that not all will have been included.

In the recording of each sequence, the deaths of the husband/father and wife mother are distinguished from each other, but the same is not true of the various marriages or deaths of the children. In other words no account is taken, for example, of the sex or the birth order of the children. Thus in the two child family the sequence is considered the same regardless of whether the elder or the younger child married first, but differs according to whether the husband/ father or the wife/mother died first. The possibility that the marriage of a child or the remarriage of the widowed parent may be subsequently terminated by divorce or the death of the spouse prior to the death of the surviving parent is not taken into account. Similarly there is no record of children 'leaving home' or of the occupational retirement of the husband/father.

The following are the distinguished events:

1. the marriage of a child;
2. the death of a child;
3. the death of the husband/father;
4. the death of the wife/mother;
5. the marriage of the widowed parent.

In addition sequences are distinguished according to the number of surviving unmarried children at the time of the death of the widowed parent. The 579 are distributed amongst 214 different sequences. These together with frequencies are appended to this paper in full. The most frequent sequence is the marriage of an only child followed by the father's death followed by the mother's death, occuring 41 times.

ANALYSIS

The main objective of the analysis is to assess the appropriateness of the family life cycle as a representation of actual family processes. The realisation of the classic family life cycle could be identified by the sequence: marriage of a child, marriage of a child, etc., up to marriage of last remaining unmarried child, death of husband/father, death of wife/mother. The number of instances of a widow's death being preceded by that of her husband which in turn was preceded by the marriages of all their children, all the children surviving to mourn her death is 94. This includes the 41 such families with only one child.

This specification of the family life cycle is perhaps a little narrow and it is perhaps surprising that as many as 94 cases were recorded. They amount to 16.2 per cent of the 579 and approximately 10.1 per cent of all the 934 families in the sample. A wider specification of the family life cycle might be that all children in a family either marry or die before the first death of a parent. Given this there will be a period which can be approximately described as the 'empty nest' which, regardless of how the children have left, leaves the two parents as the sole remaining members of the family. This specification reflects the typical assumptions that are made about family life and old age: that the ageing couple enter the final phase having seen all their children launched into families of their own. It is not stipulated that the husband subsequently dies first nor that the bereaved parent does not remarry. Only 194 families, however, conform to this specification of the family life cycle. In as many as 219 families, a parental death is the first event in the sequence. Of these it is the loss of the husband/father which predominates: there were 160 widows, of whom 26 remarried. Almost half - 28 - of the 59 husbands who were left with dependent children subsequently remarried.

Another important set - overlapping with the 219 - which fails to meet the wider specification of the family life cycle given above is the 189 families in which at least one child survives unmarried after the death of the last parent. It is reasonable to suppose that many of these children had previously left the domestic setting of the family in much the same way as their married siblings had left. If this is the case, then the domestic reality of the empty nest may have been experienced by a majority of those families with children, but some of the 'spin-offs' of this phase such as grandchildren will not have followed and in particular the family cycle will not have 'cycled'.

In 256 families the child launching phase extended into widowhood with a widowed parent attending the marriage of a child.

I feel that it can be argued that, in regard to the experiences of the individual families in this sample, the family life cycle as described in much of the literature on the sociology of the family cannot be considered to be representative. As the above statistics show, only 65 per cent of the sampled marriages both produced children and avoided divorce and only 22 per cent resulted in an empty nest period.

DISCUSSION

The reason why the family life cycle seems so acceptable is partly that most departures from the standard cycle occur later than earlier in the life of the family, and partly because the cross-sectional view of the family so frequently confirms the apparent validity of the life cycle pattern. In other words the cross-sectional survey reveals 'young' couples without children or with children, 'middle-aged' couples with growing children or adolescent children, 'late middle-aged' couples with no dependent children, elderly couples and widowed individuals living on their own. These *categories* follow a 'natural' sequence and it is presumed that since they are so common most families will pass successively through each category in the sequence. Thus no account is taken of the fact that 'late middle-aged' couples with no dependent children will include (i) couples who have never had children and (ii) couples who have only recently married, one or both of them having previously been widowed. Similarly little account is taken of the fact that in proportion of younger families with dependent children one of the parents will be a step-parent due to remarriage following early widowhood or divorce, that a proportion will have lost children who have died and that a proportion will have become parents early and that their older children will have already left home.

If we are to seriously attempt to study in detail the dynamics of whole family processes and their effects upon social status and behaviour, then it will be necessary to abandon the family life cycle in favour of some more appropriate frame of reference. In the first instance this should provide a clear specification of *the current members of a studied family* and *the relationships between them.* In particular this should be unambiguous in regard to such complex situations as step-parents, step-siblings, elderly parents living with the family, divorced parents, never-married adults, young adult children living away from home, grandparents raising grandchildren in the absence of parents, elderly sisters living together and so on. In addition it should allow for quasi-familial relationships, thereby permitting a degree of self-definition by the family. A reasonable distinction which would encompass most conceptions of the constitution of a family would be between persons who are born into the family and those who enter it as adults. Similarly another reasonable postulate is perhaps that a person only remains a member of a family for a single unbroken period of time: that no one can 'leave' a family to rejoin it at a later date.

The second element of the frame of reference should be an appreciation of the significance of *past events in the family's* history. As suggested above, the ordinary family consisting of a middle-aged couple and growing children could have had a number of events occur in the recent past: child birth, child death, child marriage or parental (re?) marriage. Indeed the couple might not necessarily be married. They might be co-habiting, possibly awaiting a divorce to come through. Given that any point in time will be preceded by a sequence of such events, the current 'stage' of the family as perceived by its members will be strongly related to the last birth, marriage, death or divorce to have occurred within it. All of these possible events are significant in that they mark the entry of exit of particular members of the family.

The third element is a little more diffuse, but could nevertheless be critical to an understanding of a family, its current circumstances and its multiple perceptions of its history and development. This is the set of all members of the family, past and present, and more generally the set of all persons with whom current members of a family have shared past family membership. For example, many grandparents exert an influence over families long past their own demises. Children who have left to marry are a critical component in the life of the family that remains. Previous spouses, and in particular children of earlier marriages, are important determinants of the routines of the married couple in a family.

In studying the *contraction* of a family defined according to the above considerations, it is reasonable to disregard entries of individual members into a family and to suggest that the first and last departures of any meaningful category will be seen to be particularly significant events. Not only will there be the individual significance of the occasion distinguishing it from intervening events, but such events will be interpreted as 'benchmarks' by many participants (Roth, 1963): events which demarcate distinct phases in the family history from the perspective of the members of that family.

It is undoubtedly true that the marriage of the last remaining unmarried child and the death of the first adult are the two most significant events in the contractions of most families. Even in the early stages of family life, it is these two events which are anticipated and for which various kinds of plans or financial provisions must be made. Nevertheless family life in old age is coloured by many other events. A set of brothers and sisters share very distinct feelings when the first of their number dies. It may well be this event rather than the loss of spouse which brings the ageing person face to face with the reality of old age. Similarly it may be the death of a last remaining uncle or aunt - the last of a generation - which is seen to be significant. When a widow remarries, then she sets a precedent for other members of her family, past and present;

a precedent which may not be followed, but if nothing else an assertion that she refuses to see her loss to represent retirement from expanding family life. Finally, this frame of reference directs attention to a previously neglected circumstance: that of orphanhood. The status of 'orphan' is normally associated with childhood, but nothing happens as one moves into adulthood which necessarily reduces the sense of loss associated with the death of parents. The fact that the event of the death of the surviving parent is not singled out for special attention by caring agencies or psychological research, does not mean that it is not an event of considerable personal significance.

This discussion is intended to suggest approaches to the study of the family which are more flexible and thereby more general, but which are also just as useful and usable empirically. At the very least a dynamic approach to the study of family life should accomodate variations in sequences of stages. Consequently one which is based upon the specification of a set of events which are free to occur in any order rather than upon a set of sequential stages or categories is clearly the more appropriate.

ABSTRACT

This paper adopts a biographical approach to the empirical study of the contraction of nuclear families. It documents and briefly analyses the frequencies in which different sequences of events occurred in a sample of 934 families. The sample is described and critically discussed. The events considered are marriages and deaths of family members. It is found that only 22 per cent conformed to a fairly broad specification of the family life cycle as presented in many family sociology texts. It is argued that this implies that the family life cycle cannot be considered to be empirically valid. Reasons for its popularity are considered and an alternative approach developed.

REFERENCES

Burke, J., *Burke's Peerage, Baronetage and Knightage*, Burke's Peerage London, 1970.

Bytheway, W.R., Sampling biases and family units, *SRUS Occasional Paper* 3, Keele, England, University of Keele, 1973.

Bytheway, W.R., Problems of Representation in the 'Three Generation Family Study', *Journal of Marriage and the Family*, 39.2, 1977, no. 2, 243-251.

Bytheway, W.R., Ageing and Sociological Studies of the family, Colloquium Paper, Ystad, Sweden, 1978.

Hollingsworth, T.H., The demography of the British Peerage, *Supplement to Population Studies*, 18, 1964, no. 2.

Johnson, M.L. That was your life: A biographical approach to later life, *Dependency or Interdependency in Old Age*, (editors: Munnichs, J.M.A. and W.J.A. van den Heuvel), Martinus Nyhoff, The Hague, 1976.

Roth, J.A. *Timetables*, Bobbs-Merrill, Indianapolis, 1963.

APPENDIX: <u>The frequencies of different sequences of events</u>

The number on the left is the frequency of occurrence of the husband/father dying after the wife/mother; the number on the right is the frequency of the same sequence but with the wife/mother dying first.

M = a child's first marriage
D = the death of a child
F = the first death of a parent
L = the death of the surviving parent
R = the remarriage of the widowed parent
X = the ever-married child alive at the time of event L

10	MFL	41		DFLX	3		FDDLX	1		DDMMFL	1
3	MFRL		1	DDFL			DMMFL	7		DDMMFDL	1
1	MFDL	3				1	DDMFRL				
3	MDFL	3	2	MMMFL	17		DMMFDDL	1	1	MMMMMFL	2
5	FML	26		MMMFRL	1		DMFML	2		MMMMMFDDDL	1
2	FMRL		1	MMMDFL	1		DMFLX	2	1	MMMMFMRL	
	FMDL	4	3	MMFML	6		DFMML	1		MMMMFLX	1
11	FLX	13		MMFMDL	1		DFMLX	1		MMMMDFL	1
10	FRML	9	5	MMFLX	8					MMMMDFLX	1
1	FRLX		1	MMFRLX		4	MMMMFL	4		MMMFMLX	1
1	FRLD			MMFDMDL	1		MMMMFDL	1		MMMFLXX	1
	FDL	4	1	MMFDL	2		MMMMDFDL	1		MMMDFLX	1
1	DFL	3	1	MMFDLX		1	MMMFML	1		MMFMMML	1
1	DFRL			MMFDDL	1		MMMFMDL	1	1	MMFMLXX	
			1	MMDFDL			MMMFLX	3	1	MMFRMMLX	
10	MMFL	30		MMDDFL	1	1	MMMFRML			MMDFMLX	1
2	MMFRL		2	MFMML	3	1	MMMDMFL			MFMMMML	1
	MMFDL	2		MFMMDL	3		MMFMML	1		MFMDLXXX	1
1	MMFDRL		2	MFMLX	6		MMFMLX	2		MFDMMMLX	1
1	MMDFL	3	1	MFMRML			MMFMDLX	1		MDMMFLX	1
	MMDDFL	1		MFMDL	1	1	MMFLXX	6	1	FMMMMML	1
2	MFML	10		MFMDLX	1		MMDMFL	1	1	FMMMMLX	1
1	MFMDL			MFLXX	3		MMDFLX	1		FMMMLXX	2
5	MFLX	13	1	MFRMML			MMDDFL	1		FMMMDML	1
	MFRML	1		MFDLXX	1	1	MFMMML		1	FMRMMMML	
1	MFDML		1	MDMFL			MFMMLX	1		DMMMMFL	1
	MFDDL	1	1	MDFML	1		MFLXXX	1		DMMMMDFL	1
	MDMFL	1		MDFLX	2		MDMMFL	1		DMMFLXX	1
	MDFL	2		MDFLXX	2		FMMMML	1		DMMDMFDL	1
	MDFLX	1	3	FMMML	7		FMMMLX	3		DFMDLXX	1
	FMML	22	1	FMMMDL	3	1	FMMLXX	1		DDMMFLX	1
	FMLX	12	2	FMMLX	5		FMMDMDL	1			
	FMRDL	1		FMMDML	1		FMMDLX	1		MMMMMFLX	1
1	FMRLX	1		EMMDLX	1		FMDLXX	1	1	MMMMMDFL	
	FMDL	4	2	FMLXX	1	1	FRMMMML	1		MMMMFLXXX	1
2	FLXX	8		FMDMML	1		FDMDML	1		MMMDMDDFLX	1
3	FRMML	6		FMDML	2	1	DMMMFL			MFMMMMLX	1
2	FRMLX	1		FMDMRL	1	1	DMMMFRL			MDFMDLXX	1
1	FRLXX	2		FMDLXX	1		DMMFML	1		DMMMFLXX	1
	FRDLX	1		FMRMLX	1		DMMFLX	2			
1	FDML	2	1	FLXXX		1	DMMFRLX		1	MMMMMMFLX	1
	FDLX	1	2	FRMMML			DMFMMDL	1		MMMDMMFLXX	1
	DMFL	8	2	FRMMLX			DMFMLX	1			
	DMFDL	1		FRMMDMDL	1		DMDMFL	1	1	MMMMFMMMMDDLX	
	DFML	3	1	FRMLXX			DFMMML	1		FMMMMMDMMDMMLX	1
	DFMRL	2		FRMDDL	1		DFLXXX	1			

Relations inside the family

4. Some comments on the relationship between aged parents and their adult children

Hannah Weihl (*)

The relationship between the two generations is obviously patterned and shaped in the course of the many years of mutual involvement with each other. This developmental process does not occur in a social and physical vacuum, but is subject to numerous constraints stemming from both these domains. Amongst the factors influencing this relationship over time are: the normative system ("culture"), level of education (of both generations), level of income or standard of living at various stage of the family life cycle, age (especially differences in age at given stages of the cycle), sex, meaningful political events; and then there are personality factors, genetic factors and physiological ones. A series of interrelated hypothesis concerning the differential influence of all such variables on the intergenerational relations in old age should be developed taking into account the impact of such variables at previous stages of the life cycle. This would be a formidable task which I, for one, certainly do not feel up to. All I intend to do at this stage is to illustrate the importance of this approach to the understanding of the pattern of intergenerational relations by posing a few questions aimed at highlighting the issue.

a. What is the influence of differences in age between the generations on the relations between them at different stages of the family life cycle, and especially in old age? Does a big or small age gap enhance supporting

(*) Brockdale Institute of Gerontology and Adult Human Development in Israel, Jerusalem.

relations in old age? How do other variables, such as culture, class, intelligence, income, fit into this discussion?

b. Does level of education, or the gap between levels of education of the two generations, influence the relationship between the generations? At what stages or points in the family life cycle, does it affect the relations in old age? How does this influence, if it exists, interrelate with other independent variables?

c. What is the influence of intergenerational mobility on the relations in old age, in different class structures or cultural settings?

d. Is there any difference in the relations between an aged parent and an adult son or an adult daughter; between married or unmarried children? Do such relations change with the change in marital status of anyone involved? What is the development of such a relationship in the course of the life cycle? How are these relationships influenced by variables such as culture, income, level of education, intelligence?

The point I have been trying to make so far is the following: intergenerational relations do develop (i.e. undergo changes); this development begins at childhood and continues throughout one's lifetime; it is patterned by numerous constraints - societal, psychological and physical ones; and I hypothesize that careful research and analysis will eventually lead towards identification of specific phases and stages in this development.

This line of thinking is obviously based on the developmental approach to family analysis. It therefore seems appropriate to point out that developmental family theory has so far disregarded the fact that the children remain part of the older couple's family even though they are grown up and mostly have their own family.

Research I have been engaged in during the last years has yielded some indicative information, mostly concerning constraints upon the intergenerational relations, but also indicating some possible lines of development.

First, and briefly, some remarks concerning factors (or constraints) influencing relations between the generations. We have looked at a small number of independent variables, the most important of which are: sex, age, income, level of formal education and cultural origin, in relation to some aspects of intergenerational relations of a population of persons aged seventy or more years. The data show that sex, income and level of formal education are of significant importance; men and women differ in their pattern of contact with children; persons whose income and education are low have different patterns of conduct with children than those who are high in both these aspects. Age of both generations seems to be of importance: the higher the age of both parents and adult children, the more they are likely to live together with satisfaction and mutual tolerance. We have also obtained interesting data on the importance of cultural differences. Israel offers a unique opportunity to study this factor, because of its being a country of immigrants. Contradicting expectations, and also contradicting myths about traditional family organization, we found that culture of origin (defined empirically as either western, i.e., person originating from Europe, or traditional, i.e., persons originating from various near eastern countries) is probably only a minor factor in patterning relations between the generations. When various aspects of such relations (visits with and by children, satisfaction with living arrangement) were examined, it was found that age of both generations, subjective health rating, income of the elderly and level of education, each of them or sometimes all of them, explain more of the variation than cultural origin. It thus seems that, at least in old age, the cultural boundaries are of little importance. We do not know if at early stages of the life cycle they had been of more importance.

There are two points I wish to make on possible development:

a. We found harldy any evidence on moving in with children,

or joining households, in the course of five years between two stages of the interviewing of the same population. Also, from a general question to all those living with children, we found very few who had ever lived apart from them. It thus seems that, at least in Israel and not culturally dependent, the pattern of living arrangement in old age is one of the two: either one remains with at least one child until death, or one separates at a comparatively early age (when the children are grown up) without really having a way back. Speaking developmentally, one may well assume that there are distinct stages in each of these lines, such as the moment one becomes physically more dependent on help from others, which mainly means that one becomes more dependent on children. This constitutes a change in the relationship with the children and thus is a phase in the relationship development.

b. We found that high age of both parent and adult child predict satisfaction with the joint household. The "children" over 40 expressed more satisfaction with this arrangement, though they also stated that living with an aged parent is a burden. Parallel to this, satisfaction with this living arrangement rises with the age of the parent. These findings very likely indicate a development over time, though one ought to remember that they are a result of cross-sectional analysis and are not based on observation over time.

In spite of the limitations mentioned, these findings do indicate that at least some of the theoretical deliberations can be developed and empirically verified. I feel that it would be worthwhile to continue this line of investigation first by developing theory and the conceptual framework, followed by some empirical probing.

SUMMARY

This short paper deals with some of the aspects of intergenerational relations in old age as a developmental process, and as part of the family life cycle. Research data indicate the important impact of various non familial variables on the patterning of such relationships.

5. Family and neighbourly relations - their role for the elderly

Halina Worach-Kardas (*)

EXISTENCE OF MULTIGENERATIONAL FAMILIES

In all industrialized countries we can observe a process of the disappearance of multigeneration families. This process has also encompassed Poland, where a small core family becomes more and more popular especially in urban areas. There exist, however, factors - mainly of cultural and traditional nature - which promote further existence of multigeneration families. A random survey conducted in Poland in 1967 among old people ranging from the age of 65 and over revealed that 67 per cent of them lived together with their children (Piotrowsky, 1973). For other countries the corresponding figures are: 42 per cent for the United Kingdom, 28 per cent for the United States and 20 per cent for Denmark (Shanas et al., 1968). Living together with children occurs more frequently in rural than in urban areas where traditional family-farms are prevalent. In the town it often happens that newly married couples live for some years with their parents who provide them with material assistance and care for their children. The other reason for joint living is that children accept an old parent (mother/father) into their house to benefit from their help in maintaining the household or to provide for parents in need.

(*) University of Lódż (Poland).

ASSISTANCE: TWO DIRECTIONS

Assistance within the framework of the multigeneration family may take two directions: from old people to children and vice versa. The assume that the help and services rendered by members of the older generation toward the younger often exceed that provided by the younger generation toward their elders. Even the principle of mutual help is accepted by both generations; in most cases assistance is of a one-sided character i.e., it is provided by the older generation. Older people prefer "giving" to "taking" and they are reluctant to accept assistance from their children (primarily material help) for fear of jeopardizing their mutual relations. The situation changes only in extreme old age when the opposite becomes true. At this stage there is a growing demand for personal care, daily services, as well as material aid. Services of this type may be provided by the social welfare system, but in many cases they are rendered by the family. In Poland - unlike in many other countries - the system of special institutions for the elderly has never been very popular. Research has shown that old people in Poland prefer small groups and informal help to the more institutionalized forms.

The family constitutes a primary, natural, universal group since within its framework various needs, not only primary but those of a higher order (according to Maslow's hierarchy) are being satisfied. Thus the family role for its senior members cannot be reduced to its protective functions. It also performs an equally important function as an "emotional safety institution". This particularly concerns the satisfaction of higher needs : acceptance, recognition, usefulness, safety. Some of the above mentioned needs may be satisfied within different small groups (e.g., circle of friends) but in most cases they are satisfied within the family.

On the other hand, traditional relationships between generations have undergone substantial change and modification. Care and support provided for the elderly by their children is no longer a simple *function* of family relationship as formerly, but it depends primarily on *personal* relations linking an old person with his or her younger relatives. Even in a situation in which these relations are fairly good the family can provide protection for the elderly only to a decreasing extent. That is due to certain unfavourable operative factors which hamper the fulfilment of the protective duties of middle-aged families in relation to their elders. They include: growing professional activity of young women, spatial and vertical mobility of young people and the common drive towards the separate housing of elderly and their children. These factors account for the decrease in the scope of protective functions of the family in relation to its senior members. On the other hand, there exist in society factors - conditioned mainly by culture and tradition - which promote the further existence of multigenerational families and intergenerational relationships in Poland.

RESEARCH DATA

Our research was limited to a sample of 278 retired manual workers in the big industrial town of Lodz. The main emphasis was placed on their social and family position and requirements in the field of social protection. The sample under survey comprised 154 women and 124 men, aged 60 and over. The family situation of the persons investigated is as follow: 58 per cent live with a spouse or with the spouse and other related persons. In total 36 per cent live together with children or with children and grandchildren i.e., within a multigeneration family. 76 per cent of the respondents had children. When questioned "Do children and grandchildren approach you with their troubles and ask for help and advice?" 75 per cent responded "yes" and 23 per cent answered "no". The most frequent forms of help provided by the aged include sharing experience, looking after grandchildren, financial help, household assistance: running the house, cooking meals, shopping, etc. To the question, "Do you approach children with your troubles, ask their advice for help?" 53 per cent answered "yes" and 44 per cent responded "no". Assistance provided by children and grandchildren comprises mainly: protection in illness, financial aid, help with housework, arranging various matters (e.g. in public offices).

Although mutual assistance links older persons with the middle-aged family members, old parents more frequently assist their children. The existence of such a "family support system" is of great importance to the elderly. What matters is not the extent of the assistance but the fact that potentially it exists.

To the question concerning whom the respondent may rely on most of all and whom she/he would approach first of all for help and advice - 56 per cent answered that they would approach children and grandchildren, 13 per cent further relatives, and 10 per cent a spouse. Thus, taken together three fourths of respondents point to their family as the first "institution" expected to provide help and advice in need.

Neighbours rank second on the list of those who offer assistance to retired persons in need. To the question whether the elderly have contacts with neighbours, 76 per cent answered positively. Thus, three fourths of those examined maintain interactions with neighbours, further responses show that these are relatively close relations. That is not an accidental fact since the examined persons have been inhabiting the same houses for several decades. They know the other inhabitants and their children quite well. Spatial proximity is here combined with social proximity. In such a working-class environment people are on friendly terms and are willing to render assistance to one another. This type of social relations, which is gradually disappearing today, prevails in former industrial settlements, and that is where a part of our respondents lived.

Today, such typical working-class housing districts are gradually being demolished and their inhabitants move to modern blocks of flats equipped with all conveniences. Although that is objectively favourable, elderly persons are often unwilling to move to modern housing giants. That is due to the fact that although their living conditions become improved, this change is synonymous to deterioration of conditions which assure potential aid and protection. Close contacts with environment and the high proportion of mutual help relations seem to account for the importance of the neighbouring support system for respondents.

Listing those who provided help and advice in need, respondents mention their own work establishment in the third place (before social welfare institutions). At the larger industrial enterprises clubs organized by trade unions and welfare departments function for the benefits of retired persons. They are centres of active cultural and social participation and also provide aid and mutual help. This care and interest of the industrial enterprise is looked upon as a "natural" continuation of the benefits to which employers are entitled during the course of their professional engagement. Our research proves that links with the work establishment are of great significance for the retired. Undoubtedly, charging enterprises which provide welfare benefits for retired personnel (are a fairly) interesting solution to be commonly adopted along with and independent of family help and social welfare institutions.

Research shows that the elderly persons have preference for such forms of help which combine care itself with social contacts, satisfy the need for social participation and provide a sense of security. This indicates that it is not merely a question of help alone but also of the satisfaction of their social and emotional needs (Rhee, 1974). This assistance should, moreover, proceed from the basic living environment of the elderly i.e., first of all from the family, the neighbouring circle and the work environment.

The research findings presented above testify to the *preferences* of the elderly. One might think that the structure of the support *really* received by them is somewhat different. Nonetheless, it is precisely these preferences which are of interest to us. The group under survey does not point to the habit of applying to social welfare institutions for help. Their reluctance in asking strangers for help and advise is promoted by ambitions and a kind of "shame". On the other hand, one can observe the tendency of approaching the family and neighbours with these problems. Research has proven that the elderly would apply for professional help only when all other sources have failed them. It would seem, nevertheless, that this phenomenon is not specific only to older people in Poland (Ciuca, 1977).

On the contrary, the traditional multigeneration family has been undergoing transformations. Possibilities of extending protection to an old mother or father are shrinking. The answer to the problem must be sought in the changing role of women in society and the need to reconcile their family responsibilities with their professional activity. This family-circle transformation accounts for the growing need for institutional assistance. The social welfare system is expected to fill this gap.

There is a view that an increase in the number of institutions and services for older people might successfully solve the problems of this group. From this point of view the social welfare system could provide basic and sufficient care for the elderly. Its big advantage is that on principle it has a universal application i.e., all persons fulfilling some conditions (e.g., definite age) are entitled to apply for this care. In the family, on the contrary, obligations are of a moral rather than legal character. This means that aged person and their children *may* provide reciprocal help to one another although they *are not obliged to*. Moreover, a part of the elderly (24 % in our survey) have no children while a part of them do not maintain close and frequent contacts with them. Thus family help and support is not a common principle or a common form of care.

Although expansion of social services and institutional care provided by the social welfare system has became a necessity, the possibilities which it offers are accompanied by some limitations. Generally speaking, these institutions aim chiefly at satisfying the primary, elementary needs, such as: providing lodgings, meals etc. (Aging in the Year 2000, 1975). Nevertheless they do not fulfill the higher needs which can only be satisfied within small, community type groups.

For the majority of our respondents it is the family which performs the important role of support system, source of personal relationship and emotional community. To the question, "What do you consider most important in life and what do you still hope to achieve" the response was again family oriented. Fifty-two per cent of the respondents played an active role in family life providing help and advice to children, caring for grandchildren etc. Twenty-four per cent considered rest and health care most important. Also these results confirm the importance of family relationships to the elderly.

CONCLUSION

It seems that without giving up all the benefits which the social welfare system provides, we should try to "rediscover" the family as well as the other small groups as a "natural support system" and field of activity for the aged. Thus the problem might be presented as follows: in setting up or improving institutions and services for the elderly we must attempt to protect the old people's social environment i.e., family, circle of friends, local community etc. These primary groups help to protect the elderly against loneliness and isolation. For society, as such, there remains the fact that family help, assistance rendered by neighbours, or just any help within one's environment is the cheapest kind of aid. Groups of the community type, based on informal ties and mutual help, are in the position to solve many life problems of the elderly and especially those of personal and emotional nature. The family is of great importance for them as the first and basic support system and emotional safety institution.

ABSTRACT

The purpose of this study was to investigate the relationships between old persons and their social environment. The subjects were 272 retired workers in the large industrial Polish town of Lodz; 36 per cent of them lived together with adult children and/or grandchildren i.e., within a multigenerational family. Three fourhts of the respondents point to their family as the first institution which is expected to provide help in situations of real necessity. Even the principle of mutual help is accepted both by older and younger generations - help and services are more often provided by older persons for children and grandchildren than the other way round. The situation changes only during extreme old age. Neighbours rank second among the list of those who offer help and advice. Seventy-six per cent of the respondents admitted they had contacts with neighbours who, in most cases, are relatively close relations. The group under survey did not indicate the existence of the habit of applying to social welfare institutions for help or advice. On the other hand, there can be observed the custom of approaching the family and the other small groups with these problems. The results of the research confirm the importance of family relationships among the elderly in Poland.

REFERENCES

Aging in the Year 2000, A Look at the Future, *Gerontologist*, 15, 1975, no. 1, part II.

Ciuca, A., The Elderly and The Family, paper presented at the Symposium of the European Social Sciences Research Committee, Ystad, September, 1977.

Piotrowski, J., *Miejsce czlowieka starego w rodzinie i spoleczenstwie (Old People in Family and Society)*, Warsaw, 1973.

Rhee, H.A., *Human Aging and Retirement*, General Secretariat, International Social Security Association, Geneva, 1974.

Shanas, E., P. Townsend, D.Wedderburn, H. Friis, P. Milhøj, J. Stehouwer, *Old People in Three Industrial Societies*, Routledge and Kegan Paul, London, New York, 1968.

Worach-Kardas, H., *Nauczyciele a emerytura (Teachers and Retirement)* Warsaw, 1973.

Worach-Kardas, H., Institutional Care-Possibilities and Limits, *Proceedings for the XVIII th Congress of Social Gerontology*, Krakow, Poland, 1978.

6. The elderly and the family

Alexandru Ciucă (*)

INTRODUCTION

Within the framework of the interdisciplinary research carried out at the Research Department of Social Gerontology of The National Institute of Gerontology and Geriatrics, Bucharest, a series of problems related to the elderly and the family have been studied, such as: the role and status of the elderly in the family, the characteristics of family relationships and of intergenerational relationships, the correlations between different indicators of family context, of health status and of adjustment.

Some of the results of these researches confirmed the characteristics already discovered in other countries. For instance, the trend toward a nuclear-type family was conspicuous in the urban area, as opposed to the rural area, where the percentage of elderly living with their adult offspring is significantly higher.

THE DATA

The studies carried out on a representative sample of 3,000 subjects of 14 districts of the country, pointed to the fact that in the rural area 50 per cent of the elderly are living with their offspring, while in the urban area, the proportion is 10 to 30 per cent.

Owing to the high rate of industrialization and urbani-

(*) National Institute of Gerontology and Geriatric, Bucarest.

zation, characteristic situations are to be found in the urban areas of Romania as regards the integration of the elderly into the family. In the new industrial towns built up in recent years, the population is for the most part young (coming from rural areas where aged parents were left behind) and the tendency to create biological families prevails. On the contrary, in the century-old or decades-old towns the demographic aging phenomenon, correlated to the integration of the aged persons into the families of adult offspring, is still present.

The elderly continue to maintain relationships with their offspring even when they no longer share the same household. These relationships consist of mutual help, visits, phone calls etc.; 70 per cent of the surveyed subjects stated that these kinds of relationships are very satisfactory. Since the industrialization and urbanization process is relatively recent in Romania, the intergenerational relations are still very strong and similar in many respects to those of the rural area.

One of the main objectives of this study was to find out how the interrelations were perceived and their significance to different generations.

Potentially, the family offers the elderly favourable conditions of a "protected independence", with the advantages of affective relations and a sense of security. The need of security increases with age, and cohabitation with offspring or at least with relatives is highly desirable. A survey carried out on a sample of 3,000 subjects, aged 56 to 90, pointed to a significant rise in the proportion of those who wish to cohabit with their offspring in the most advanced decade of their life.

A noticeable fact among all age groups was that the percentage of those who wished to be admitted to an institution for the elderly (home, hospital-home etc.) was the lowest. This is due to their apprehension of possible unsatisfactory material conditions and especially to their rejection of the idea of dependence upon strangers, which

Table 1. Preference concerning household arrangements (percentage distribution) N = 5000

Age group	per cent with off-spring	per cent separate-ly	per cent in an in-stitution for elder-ly	per cent with rela-tives	per cent refusing an answer
+ 50	30.5	35.5	5.1	15.9	13
60-69	29.5	38.7	5.2	12.2	14.4
70-79	26.3	36.9	6.9	14.3	15.6
80 +	46.4	23.6	7.1	12.5	5.4

implies a lack of emotional ties or moral obligations.

The characteristics of intergenerational relations and of the elderly's role in the common household were investigated through intensive interviews of 2,500 adults and aged over 65, living together at home. In the case of 80 per cent of the middle aged couples both spouses were employed, so that, cohabiting was based on mutual interests and the aged played an active role in the family. The initial hypothesis that the elderly had an especially active role only in bringing up and taking care of grandchildren is obsolete; a leading role of the elderly is that of taking important decisions regarding the common life of the family members. Consequently:

over 50 per cent were household leaders;

30 per cent performed other functions;

15 per cent had no activity within the family framework.

The opinion of the middle generation with respect to the role of the elderly in the family is in accordance with facts and when quite favourable as the following data indicate. Elderly folks are:

- a valuable help: 75 per cent;
- a family obligation: 21.8 per cent;
- a burden: 3.2 per cent.

The favourable opinion regarding the elderly is sometimes reinforced by certain material advantages of interest to the middle generation. This situation, was found to be true with 30 per cent of the elderly's families covered by the research.

A factor influencing negatively the intergenerational relationships is the poor health status of the aging person. The answers to the question regarding care during temporary or long-term disability were as follows:

- care with affection: 55 per cent;
- care without affection: 30 per cent;
- without care: 15 per cent.

It was then conspicuous that 45 per cent of the elderly with health-problems were not satisfied with the care they received from the other family members: however, 60 per cent admitted that they felt secure when attended in the family during illness.

The negative influence of certain degenerative diseases (irrespective of the sharing of the household) showed up in the elderly persons' qualitative assessment of the relationship with their offspring and relatives. 27.5 per cent of these subjects considered the relationships with the younger generations better than before retirement (due to more leisure and the meaningful activity carried out within the family framework); 36.6 per cent found the relationships worse and attributed it to the deteriorating status of their health.

As a result of the interviews carried out, the causes of conflicts emerging from life in common were also analysed. The elderly corroborate the answers of the middle generation regarding the right to take decisions and to be consulted for the main problems; conflictual situations emerge, however, due to different tastes (44 per cent), the different preoccupations (60 per cent) and different opinions with regard to moral values and principles of education.

The correlation between the relationship degree (including the social relations in general) are better when advancement in age is complemented by a better health status. On the contrary, the negative correlation of the relationship are more significant when advancement in age is conjoined with the deteriorating state of health.

This situation is also mirrored in the correlations between the intensity and the quality of the family relationships and the two mentioned indicators; of the social relations the family relations were found to be more intense and more durable.

Active participation in the family life is directly correlated with the elderly persons' feeling of satisfaction because it confirms the feeling of social utility.

CONCLUSION

Social relationships and especially family relationships compensate for the lack of professional activity and meet the basic human needs for security, communication and affection.

7. The influence of old people on the living standards of a family

Andrzej Tymowski (*)

The problem of an aged family member and his influence on the living standards of the family is not a new question; it is as old as mankind itself. According to the popular notion there exist two categories of old persons. The first are owners of a "property", a case especially popular in the rural areas of the sterotyped, aged farm owners; the second are those who are unable to support themselves either because they have a small or no income and must rely on the family for support - a fact which contemporary society considers burdensome. The problem as such is complex and worth a thorough analysis from various points of view.

"Old" people cannot be considered as a uniform group. The very process of "getting old" has an individual character and it is hard to say where lies the border line beyond which one enters old age. Since the definition of this border line is necessary both for statistical purposes and those of social policy one may assume that the age limit at which man is considered old is the fixed age of retirement. There are exceptions, of course, - an earlier retirement or a later one for particular professional groups, the specific situation of individual farmers or women - wives and mothers - who do not earn their living. In general, one may assume that old age begins with the moment one ceases to be a wage earner. The fixed retirement age limit corresponds on the macro scale with physiological

(*) Research Department, Institute of Home Trade and Services.

processes occurring in an aging organism, processes that are responsible for the systematic decrease of paid work capacities of man, for his lower productivity and limited physical efficiency.

According to data regarding Poland, a woman at the moment of her retirement (60 years) has on the average 19 years of life ahead of her, while a man retiring at 65 still has 12 and a half years. The period of retirement is thus relatively long and the effect which the presence of an old person has on a family household is of great social importance. To begin with, it is most essential to accept one truth. One cannot speak of some homogeneous collectivity of old people having a uniform influence on the family's living standard. On the contrary, a considerable segmentation of the collectivity is necessary. According to sex and age, as well as the fact of living alone or living with a single or multigeneration family or even confined to a rest home for the aged. Further criteria of the segmentation are: profession performed during the period of professional activity, residence - the differentiation between town and rural area is here of particular importance - and, undoubtedly the physical efficiency of the aged person, correlated with his/her age and sex. Another problem is the right to the social security annuity or the lack of it, property ownership and the standard of household equipment, two possibilities of extra paid work after the period of retirement or of work in the household. The following reflections aim at showing the logical dependence which exists between the presence of an aged person in the family household and its standard of living. People who live alone, have no family or no family ties on the material level in the broad sense of the term, are thus not considered here ex definitio.

Income and personal means i.e., the social security annuity and income from paid work after retirement as starting points of the discussion. In Poland the social security depends on the one hand on wages received in the period of active professional work (for a freely chosen period of one year) and on the other on the age of the old man's retirement. In general, the social security annuity is relatively high at the beginning of the retirement period (though of course it is much lower than the wages; in 1975 an average annuity amounted to only 43 per cent of the average wages). In the course of time, however, the real value of the social security annuity decreases due firstly to the increase of prices, not always compensated for by the increase of the annuities themselves, and secondly due to the fact that the rate of growth of nominal wages is higher than that of the social security annuities. The recently undertaken long-range reforms aiming at the improvement of the living standard of the so-called "old portfolio" social security beneficiaries, as well as a special extra allowance for those who have reached 80 years of age, are the remedial measures which can bring about some improvement in the state of affairs here referred to. In fact, only a small group of retired persons (about 1/10) has an income from paid work performed after retirement. The income received within the frames of official regulations is in principle limited by the law to a rather low level, differentiated according to the source of earnings or the fixed general limit of earnings per year. The extra earnings as well as the rate of professional activity of the retired people evidently decrease along with their getting older and after the age of 70 practically disappear on the macro scale. When estimating data concerning extra earnings one has to stress, however, that the statistical data in this respect are incomplete and not quite reliable. For reasons of taxation and due to the fact that the social security annuity may be withheld when income exceeds a certain limit, retired people tend to keep their extra earnings secret when they come from services rendered to private persons outside the frames of the nationalized economy. Typical examples of such services are: private lessons, hired children's care, hired household help such as charwomen etc.

Another problem is the standard of the household equipment of the retired people and of their property. With the exception of the old farm owners in the rural areas, the property of old people does not, in general, play any important role in the socialist economy. It goes without saying that the households of persons who received higher salaries are in most cases better equipped than those of people whose earnings were smaller. The difference, however, is of a qualitative and not of a quantitative nature. The standard of the household equipment is poorer in the households of older, retired people than in those of the younger ones. The most essential problem is the question of the legal form of the flat inhabited by the retired people and of its standard, a most vital question for an annuitant. Curiously enough, the ownership of the inhabited flat plays a less important role than the monthly payment. The time when the flat was obtained is important in estimating the living standard. In most cases, for instance, the later the cooperative flat has been acquired, the higher is the monthly payment. One fact is of great importance: the payment for flats of the same standard may differ considerably. Another question is of equal importance: retired people having larger flats and living alone are in a position to get a considerable income from subletting "in secret" a part of their flat for sums amounting to their annuity or even higher. There are, however, no reliable statistical data to substantiate this. A more essential fact and one on a rather mass scale is that a considerable proportion of retired people - about one fourth - live in flats of an unsatisfactory standard, a fact resulting from the great shortage of homes in Poland. The situation calls for changes but the possibilities of acquiring a new flat are in practice greatly limited for this group of people due mostly to their low income.

When estimating the above data one may state that in 1975 the income of one fourth to one third of the retired people was lower than the so-called social minimum or just on the borderline. One of the fundamental aims of the reform of the annuity system effected in 1977 was to do away with this state of affairs.

The standard of household equipment and of the property (restricted to very modest belongings) does not result - in the case of the majority of the retired people - in a marked difference in their living standard. People who are better-off are an exception.

The question concerning the conditions of life of the old people in the rural areas would call for a separate treatment because of the specific character of the circumstances. With regard to the rural population, however, it is worth mentioning that the problem of the individual farmers is most crucial for old people. It is associated, on one hand, with the gradual manpower shortage and, consequently, with the very high wages paid hired farmhands at the peak of the farming season. Even the owners of relatively larger farms have difficulties if they are not fully efficient themselves or have no relatives to help them.

The low income of the greater part of the retired persons and of the aged, in general, complemented by the unsatisfactory help of the social welfare agencies and the chronic shortage of homes for the aged, forces them - out of sheer necessity - to depend on family support. Their is a clear correlation between the aged person's age and sex and his family's support: the older the person, the worse his living standard and the greater the need for help from his family. The phenomenon is all the more conspicious with regard to women both because of their longer life span on the macro scale and their earlier retirement age. It is very hard to establish the range of the family support given to old people as well as that given by annuitants to their family. Both parties are interested in keeping it secret and it is often disguised under the form of 'gift'. In the case of common households shared by multi-generational families it is practically impossible to establish reciprocal material ties, the greater part of the income being used for common purposes. Considering the situation in Poland, one may risk the assumption that in the first period of the retirement the annuity is a rather important element of the family household income; the retired person pays his living over and above his dues especially if one considers the fact that usually he is the owner of the flat which he shares with his multigenerational family. In the course of years the situation changes radically: the real value of the annuity decreases, the wages of other members of the family increase and gradually the old man, dependent more or less on family support becomes a financial burden. A problem apart are annuitants living alone and keeping in touch with their families. An examination of the family budget points to some regularities: services in the form of gifts are usually considered of greater value when the recipient is a third party rather than an aged person; the aged manifest an astonishing desire to bestow gifts upon members of their family who are even better-off than they.

When one discusses the problem of family ties it is important to take into account the 'time budget' which though indirectly is none-the-less essentially associated with the living standard of the household. Although this element often passes by unnoticed, it actually is of utmost importance. An examination of the 'time budget' shows a substantial, though not uniform, commitment of retired women to keeping their children's house. They do the shopping, (which under the existing circumstances in Poland is a very time-consuming task) the housework, and often devote much time in taking care of the children. The help rendered by retired women enables their daughters or daughters-in-law to engage in professional work, eases the family chores of other members of the family, contributes to the visible improvement of the quality of life, enables those who lead an active professional life to relax at home or enjoy some diversion. It does not seem right to ignore this aspect when estimating material links between an old person and the family, since such an approach leads to a distortion of the real role of the aged and retired women in particular. Another noteworthy aspect is that frequently old people who live alone and maintain their own household also help their children with the housework.

There is, however a dark side to the picture. As the physical efficiency of the aged declines, they are unable to help with the household chores. At this stage, rather, they themselves need attention particularly during periods of long illness. Not only does this affect the cost of living but places upon the family the obligation of rendering the necessary services to the aged. It is usually impossible to get paid help because of the high fee involved and its scarce supply. The younger generation has then to devote its time and energies to the care of the old person. Very often this implies giving up regular jobs or part-time work and sacrificing entertainment or rest. Therefore, when estimating the living standard it is necessary to take into account also the relations between old people and their family from the point of view of the 'time budget'.

As has already been mentioned, relations between the aged farmers and their families are a separate problem. One has to stress that these relations are of a somewhat different nature. As a rule the farmer is not willing to turn his landed property over to his relatives during his life time in order to preserve his privileged status in the family. The mass phenomenon of the post-war period - the emigration of the young generation from rural areas to towns - has added to the complexity of the problem. There are now farms where only the aged remain. In order to assist this group of people, a rather complicated annuity system has been elaborated of late giving farmers the right to social security payments on the condition that they transfer the ownership of their landed property to the State. However it will be possible to estimate the efficiency of this system only after several years of operation.

It would seem that the most essential problems for discussion and further research are the following:

a. How is one to get real data pertaining to the living standard of the old people and their income, in particular, and how trace their material bonds with the family.
b. The 'time budget' of the aged has not yet been sufficiently discussed with regard to the services rendered to and by the family. It is a most vital question for the estimation of the material ties of the old people with their families.
c. There is a necessity of launching research on the evolution of the consumption model of the aged as they get older and of the possibilities of the realization of this model on their own or in connection with their families.

ABSTRACT

The paper discusses the following problems: Effect of the aged upon the living standard of the family as illustrated by households in Poland and the complex nature of the problem;various regularities with respect to the situation in urban and rural areas; differences depending on demographic structure of the household; physical efficiency; standard of the household equipment, as well as sources of income and its amount along with its evolution in the course of time; the importance of reciprocal relations between the 'time budget' of the aged in the family and the 'time budget' of other members of the household; the mutual and fluctuating character of the relations; specific features of rural population and problems to be analysed.

8. Family contacts and social class in the early stages of old age

Henning Olsen (*)

INTRODUCTION

The main purpose of this paper is not generally to show that the study of the elderly reveals socio-economic inequality in Denmark. It has been shown several times that class differences in the population as a whole are also to be found among elderly people (see references Olsen, 1974, 223). The present paper has the more limited aim of uncovering specific class patterns of family contacts and isolation in the early stages of old age. It is also the intention, in this connection, to elucidate whether retirement will influence the patterns of family contacts in a specific way.

The results submitted in the following originate mainly from a report by the Danish National Institute of Social Research on a longitudinal study of the elderly (Olsen, 1976). The principal subject of the report is family relations, and an analysis has been carried out of the development in interviewed elderly people's contact with their children and other relatives over a nine-year-period, at which time they were 62-64 to 71-73 years of age.[1]

(*) The Danish National Institute of Social Research, Copenhagen.

1. The group of persons included in the longitudinal study is a national random sample of about 1000 persons born in the period from the 1st of May 1897 - 30th of April 1900. Interviews were conducted with these people in 1962, 1965, 1968 and 1971. Thus, those interviewed were 62-64 years old when the first survey was conducted, and 71-73 years old when the final survey took place. Of the original sample 54 per cent have answered the questions put to them during all four interviews. The results submitted in the following are mainly based on this 54 per cent interviewed all four times.

The report deals with the development of family contacts within various groups of elderly people, including different social classes: for men and women, for people living alone compared with people living together, etc. It is also an attempt to elucidate what effects various important events in life, for instance, retirement or widowhood, will have on the extent of contacts with children and other relatives. As indicated, the following will, however, be restricted to two variables, i.e., social class affiliation and employment of the interviewed persons.

In order to determine, in some field or other, whether inequality specific to social class exists, it is necessary to determine and operationalize the concept of class. This can be done in many ways.[1] In this paper the presentation of social class will be restricted to an account of the way in which this concept was determined in the Danish longitudinal study of the elderly.

As a basic criterion for social class affiliation, ownership of the means of production should be pointed out. However, it is not sufficient only to distinguish between buyer and seller of working power. Small independent tradesmen, for instance, may well be characterized as owners of means of production, but they are not in all cases buyers of working power. On the other hand, it should be pointed out that salaries to employees holding managerial posts may include amounts which do not emanate exclusively from the skills of their own labour, but are determined by their position in the hierarchy of leaders. Thus, although in their case, it is not necessarily a question of ownership of means of production, for example in the form of shareholdings, they have, nevertheless, a right of leadership and in many situations act on equal terms with the buyer of working power. Accordingly, another objective criterion is whether the persons in the work process enter into a superior or subordinate relationship with other

1. For a further discussion of this, see: H. Olsen, 1975, 223 ff.

persons, i.e., whether or not they have a right to be in charge of the production process.

Thus, taking the work process as a starting point, there are two essential objective criteria which are decisive for a person's social position. First of all, the ownership of means of production because dependently employed persons, irrespective of their possible right of leadership, in fact occupy a position different from those in self-employment. Secondly, the right of managing or organizing the work process so that the person concerned is placed in a key position,as compared with other persons. By using the criteria thus emerging as dimensions in a grouping of persons, the following four social classes were arrived at:

1. *"Salaried employees in more responsible positions"*, not having the ownership of means of production but being in charge of subordinates. Key employees, both in the private and public sectors, for example, will come within this category.
2. *"Workers"*, who are characterized by not having any ownership and not having any subordinates. Industrial workers and salaried employees in subordinate positions, among others, are placed in this category.
3. *"Larger self-employed persons"*, who are characterized by the ownership of means of production as well as being in charge of subordinates.
4. *"Smaller self-employed persons"*, comprising persons who have the ownership of means of production but no subordinates. In this group are to be found, for example, free lance engineers and small tradesmen or small-scale farmers, such as small-holders, not employing any one or only assisted by their wives.

The question of ownership of means of production was not explicitly stated in the questionnaires. The interviewed person's "life occupation", i.e., the occupation which the person concerned states that he/she has had most of his/her life, has been used as an indicator of this.

SOCIAL CLASS, FAMILY CONTACTS, ISOLATION AND LONELINESS

In the Danish longitudinal investigation of the elderly, the development of the elderly's contacts with their children, brothers and sisters as well as other relatives, and with their family en bloc (total family contacts) has been studied. Finally, the problems of isolation and loneliness have been touched upon. The results produced will be submitted in the order just mentioned, with focus especially on differences, if any, between the two social classes differing most in a socio-economic respect, i.e., "workers" contrasted with "larger self-employed persons".

In the analysis of contacts with children, two contact scales were established, i.e., "the primary contact scale", expressing the time when the elderly person last saw one of his/her children, and "the secondary contact scale", indicating whether, within the last 12 months, the elderly person has stayed overnight with one of his/her children living away from home.[1]

With regard to the time when the elderly person last saw one of his/her children ("the primary contact scale"), it appears that over the period of investigation there is a slight downward trend in the extent of the contact between the elderly persons and their children, whereas there is no definite tendency of development for "the secondary contact scale". The development in the primary and secondary scales with regard to social class is interesting from the point of view of inequality. It appears, in the case of both contact scales, that the "larger self-employed persons", being the group best off socio-economically, have also had the most frequent contacts with children during the whole period of investigation. Thus, it seems as if persons in the so-called "higher" social classes have more frequent contacts with their children during the

1. It should be noted that, in contrast to the secondary contact scale, the primary contact scale includes elderly people who live in the same household or house of at least one of their children, or in a flat on the same floor as at least one of their children.

early stages of old age than persons from "lower" classes - and more easily will be able to retain these contacts with increasing age. The latter is particularly pronounced as far as the secondary contact scale is concerned. It appears, for example, that the proportion of "workers" who in the first and last phases (1962 and 1971) state that they have stayed overnight with their children or vice versa is unchanged, whereas in the case of "larger self-employed persons" there is an increase from the first to the last phase from 17 to 27 per cent. On the other hand, this result can hardly be surprising as the findings of other studies show similar trends (Riley, 1968, 546).

As mentioned, the elderly's contacts with brothers and sisters and other relatives were also elucidated in the investigation. When analysing these contacts, there appears to be a slight variation according to the age of the interviewed persons so that the frequency of contact decreases with age. Yet, there is no marked fall, which may be connected with the fact that only those relatives with whom the elderly person actually has contact were included in the analysis.[1] If one then looks at the development in contacts with relatives with regard to social class, it appears that "workers" have more frequent contacts during the whole period than the other three social classes. However, the frequency of the contact which "workers" have with relatives tends to dwindle with time, whereas in the case of "larger self-employed persons", the frequency shows an upward trend.

The elderly people's contacts with children and other relatives have also been considered together. Two contact scales for total family contacts (TFC_1 and TFC_2) were established. The former includes "primary contacts" with children and contacts with relatives, whereas the latter in addition to this includes "secondary contacts" with children. As far as both contact scales are concerned, the

1. In contrast to contacts with children, where all children who are alive have been included.

basis group from which the percentages are calculated in this case also includes persons without children and/or relatives. The reasons for this are of course that what is now to be elucidated is the total family contacts, which implies that, for example, the group of people having no relatives with whom they associate must be relevant from the point of view of contact.

As there is a certain downward trend throughout the whole period of investigation - both in the extent of the elderly's contacts with children and with relatives - it is not surprising that the total family contacts show the same tendency. The relative proportion of interviewed persons who have seen both a child and a relative within the past week drops from 30 per cent in 1965 to 22 per cent in 1971. Correspondingly, the proportion of interviewed persons who have had contact with neither children nor relatives within the past week rises, i.e., from 14 to 18 per cent (TFC_1). As far as the development in TFC_2 is concerned, it is found that here, too, there is a drop in the contacts, as, in 1965, 16 per cent have seen both a relative and a child within the past week and stayed overnight with a child within the last 12 months, against 12 per cent in 1968 and 9 per cent in 1971.

As regards TFC development patterns specific to social class, such patterns emerge from the data material. In the 1965-phase, there is a slight tendency for "workers" to have more frequent total family contacts than the "larger self-employed persons". But it is remarkable that the picture changes in the following years, the frequency of "workers'" contacts falling markedly in the next two phases, whereas the "larger self-employed persons'" contacts seem to increase. While 18 per cent of the "larger self-employed persons" had had contact with neither children nor relatives within the past week in the 1965-phase, the corresponding proportion was 13 per cent in the 1971-phase, whereas for "workers" the corresponding proportions were 14 and 21 per cent. The development in TFC_2 further confirms the above-mentioned tendency. Thus, it seems as if there exist social mechanisms enabling persons from "higher" social classes more easily to retain and develop family contacts with increasing age.

It is a limitation of the longitudinal investigation of the elderly that it is restricted to elucidating quantitative aspects of their family contacts. This will, naturally, cause uncertainty as to the importance of family contacts, as it is impossible either to estimate how large a proportion family contacts constitute of the total social contacts, or to obtain any knowledge of the interviewed person's evaluation of the importance of his/her contact with children and other relatives. However, in 1962 and 1971 the interviewed persons were asked whether they were often alone and whether they ever felt lonely. The answers to these questions enable us indirectly to clarify the limitations referred to above.

By asking the question: "Are you often alone?" the intention was to get an insight into the elderly persons' feeling of *isolation*. It appears that the proportion of people feeling isolated increases with age. 15 per cent answered "often alone" at the age of 62-64, as against 26 per cent when aged 71-73. If, then, the interviewed persons' social class affiliation is again considered, there is a distinction between "workers" and "self-employed persons". Of "workers" about one quarter are "often alone" when 62-64 years old, and this proportion is, generally speaking, unchanged when they are aged 71-73. On the other hand, few of the self-employed persons" (about 7 per cent) are "often alone" at the age of 62-64, but the proportion rises till they reach the age of 71-73, without however reaching the level of "workers". As the "self-employed persons" are known to retire from active employment at a later age, this difference indicates that retirement may have an influence on how often the interviewed person is alone.

With regard to the responses to the question of *loneliness*, there is also a variation according to age. Among those aged 62-64, 2 per cent state that they "often feel lonely". When aged 71-73, the same categories have risen to 7 per cent and 16 per cent respectively, which means that if the two categories are considered together, there is a doubling of the proportion of lonely people over the period. If we look at the social class affiliation, this is of minor importance to the frequency of experienced loneliness, whereas retirement appears to play an essential part, resulting in a sharp increase in the degree of experienced loneliness.

After considering separately those who are often alone and those who often feel lonely, in relation to two background variables, we will take a brief look at those who are *both* often alone *and* lonely. It has often been assumed that the feeling of loneliness is most widespread among socially isolated elderly persons. It appears from the material that 5 per cent of those aged 62-64 state that they are often alone as well as lonely. This rate had tripled nine years later. Of those "often alone", 33 per cent and 53 per cent were lonely in 1962 and 1971 respectively, which, of course, implies that social isolation in the last phase more frequently made the interviewed person feel lonely. It is interesting to note that this increased frequency of loneliness is counterbalanced by a corresponding relative increase in the number of lonely persons among those who do not feel isolated, so that both at the age of 62-64 and the age of 71-73 there are just over four times as many lonely persons among those who are often alone as among those who do not feel isolated.

The differences as to social class which were mentioned in the preceding sections treating those who were isolated and those who were lonely are, broadly speaking, encountered when those who are *both* isolated *and* lonely are considered. In the group of "larger self-employed persons", very few were often alone and lonely in 1962, whereas this situation applies to 9 per cent of the "workers". In 1971, this difference is slightly reduced, but the "larger self-employed persons" are still less isolated and lonely than "workers".

Thus, if the development trends with regard to patterns of family contacts, isolation and loneliness are taken together, they seem to confirm the hypothesis of class specificity so that "larger self-employed persons" can look forward to a higher degree of family integration than "workers" in the early stages of old age.

SOCIAL CLASSES AND THE EFFECT OF RETIREMENT

The retirement process

What has been stated up to now has been within a descriptive framework, leaving the problems of effect, i.e., what happens to one variable if a change occurs in another, completely out of consideration. However, a longitudinal investigation makes it possible to elucidate effects, and in the following the influence of elderly persons' retirement from working life on the extent of family contacts, isolation and loneliness will be considered. By way of introduction, however, it seems expedient to give a brief outline of the way in which the various social classes will retire from working life (Olsen et al. 1977).

If we look at the retirement patterns for men, specific to social class, as revealed by our data, the following trends emerge: the most marked difference appears when comparing "workers" to "larger self-employed persons". While the group of male "larger self-employed persons" can look forward to what could be termed a gradual retirement, the retirement process for male "workers" is rather by jumps, nearly two-thirds of them retiring between the second and the third survey phase, i.e., around the pensionable age. It should further be noted that a comparison of the retirement patterns of "workers" and "larger self-employed persons" with four subordinates or more will accentuate the above mentioned differences in retirement. For example, it is seen in the last survey phase that there are nearly 15 times as many male "larger self-employed persons" with four subordinates or more as male "workers" who are still in active employment (N = 16 and 88 respectively). These "larger self-employed persons" are persons who can, to a great extent, arrange their own work process, effectuate their own job adaptation, reduced working hours, etc. Finally, with regard to the retirement patterns of "employees in more responsible positions" and "smaller self-employed persons", it is seen that these patterns follow a fairly continuous course if a comparison is made with the male "workers", but at the same time a jumpy course as compared with the "larger self-employed persons".[1] It is primarily "larger self-employed persons" and "salaried employees in more responsible positions" who can remain in their jobs and work shorter hours, whereas "workers" must, on the whole, find other work if they want to work less. However, most frequently "workers" must adjust to the labour market in the special way of retirement from regular employment.

Effect of retirement, specific to social class

It is a much debated question whether retirement from active employment will influence the elderly's social contacts, and, if that is the case, in what way. When, in the following, it is to be examined whether retirement will have a different effect in different social classes, the statement will be based on the two contradictory classes: "workers" and "larger self-employed persons". First, however, we will look at the effects of retirement on family contacts, etc. - without direct regard to social class affiliation.[2]

1. Regarding elderly women in active employment, it was found that they start retiring much earlier than men: thus, the relative level of women not in active employment was rather high throughout the whole period of investigation.

2. An effort was made to avoid methodological discussions in this paper, but it should be stated that when measuring the effect, the "retirement cohort method" used by G.F. Streib and C.J. Schneider (1971, 41 ff) was one of the methods used.

Regarding the importance of retirement in relationship to contacts with children, it was found that the contact of interviewed persons in active employment shows an upward trend over the period of investigation, whereas the opposite was found to be true in the case of retired persons. Moreover, during the entire period, the primary contacts with children of those constantly employed exceeded the extent of the contacts of those who had already retired at the beginning of the investigation. Although these observations do not concern the problems of effect *directly*, they are nevertheless interesting because they support the previously confirmed hypothesis of contact patterns specific to social class. For it is known that it is primarily wage-earners who are most prone to early retirement, while a large proportion of "larger self-employed persons" continues to work beyond age 67 (normal pensionable age in Denmark for men). It is therefore to be assumed that the findings are first and foremost an expression of contact patterns specific to social class in the period around retirement. As to persons retiring during the survey period, it appeared that in any case retirement will not cause an increase in the frequency of the elderly persons' contacts with their children. Nor does it seem as if retirement will result in significant changes in the frequency of contacts with relatives, but it should be mentioned that interviewed persons who have been in active employment during the whole period of investigation have more frequent contacts with relatives than persons who had already retired at the beginning of the investigation. This seems to substantiate what has previously been stated about contact patterns specific to social class.

In the light of this, it is hardly surprising that retirement was found to have no effect on the total family contacts as well. As pointed out earlier, it should, however, be emphasized that interviewed persons who were retired in all phases throughout the period have a lower frequency of family contacts than the persons who were in active employment during the whole period. This applies to both TFC_1 and TFC_2 and is connected, among other things, with the contact patterns specific to social class. Here, the conclusion can be drawn that not being in active employment does not provide an incentive for comparatively frequent family contacts; one might rather speak of a "cumulative activity effect" so that persons who are engaged in active employment also seem to be relatively more "family active".

Finally, with regard to the effect of retirement on isolation and loneliness, this can only be explained with difficulty, as data concerning isolation and loneliness are only available from the first and last phase of the investigation. In order to determine whether retirement from occupational employment will influence the percentage level of those who are "often alone", the persons who retired between 1962 and 1971 have been segregated and compared with those who either remained in active employment or were in retirement during the entire period. It was found that of those who retired in the course of the period of investigation, 16 per cent were "often alone" in 1962, and this proportion rises till 1971 when it reaches 24 per cent, which is twice the ratio of those still engaged in active employment. With regard to experienced loneliness, it appeared that only 5 per cent of the interviewed persons in active employment in 1962 were "lonely", whereas five times as many of the retired persons admitted this. For persons who remained in active employment during the nine-year period, the proportion of "lonely" ones remained at the same level. On the other hand, there was a sharp rise for retiring persons. In the latter case, the proportion of "lonely" persons rose almost five times as high, which must imply that retirement from active employment - and the loss of a number of social functions likely to be connected with this - will have a number of unwanted and unpleasant consequences. If, finally, we consider loneliness and isolation together, it appears that very few of those who were in regular employment during the whole period were *both* "alone" *and* "lonely". By contrast, there is a rise among persons who retired at some point during the period of investigation. Thus, all things considered, it does not seem as if retirement will promote family contacts, and it seems to have a negative effect on isolation and loneliness, but this is a "generalization" which does not take into account that retirement may produce different effects in different social classes. However, neither in the case of "workers" nor in the case of "larger self-employed persons" did retirement appear to produce any positive or negative effect on the extent of contacts with children. On the other hand, it appeared, in the case of both contact with relatives and total family contacts that for "workers" retirement tended to result in *reduced* contacts, whereas the opposite seems to be the case for "larger self-employed persons". Thus, the hypothesis of a retirement effect specific to class cannot be rejected, the more so as retirement will also involve a higher degree of isolation and loneliness for "workers" than for "larger self-employed persons". Therefore, the relatively higher degree of family integration in the case of "larger self-employed persons" cannot merely be considered in connection with the previously mentioned "cumulative activity effect", viz., that persons who are active in one field (employment) will also tend to be active in other fields (family contacts), but must also be viewed in connection with their relatively smaller "disintegrating" reaction to retirement. The investigation has shown that the later a person retires from work, the more frequent family contacts will he/she have, and the weaker will the feeling of loneliness and isolation be - irrespective of social class affiliation. As "larger self-employed persons" retire relatively late, this will to some extent explain the differences specific to social class, but this is only a pseudo explanation. Research work still remains to be done if one desires to uncover real causal relations which can show structures and processes that also in the field of family integration creates class inequality in old age.

SOCIO-POLITICAL CONSIDERATIONS

It is a well-known fact that the elderly do not constitute a homogeneous group, and that studies of the elderly's living conditions also reflect socio-economic inequality in Denmark. A manifestation of such inequality among the elderly in fields like economy, consumption, and housing does not imply that corresponding inequalities exist in the field which is the subject of this paper, i.e., the development of elderly people's family contacts. Nevertheless, inequality indicators have also been uncovered in this field, as the frequency of family contacts as well as the experience of isolation and loneliness could not be viewed independent of the interviewed persons' social class affiliation. The results of the investigation indicate that the contact problems connected with old age must also be viewed in connection with the class structure of society. It cannot be excluded that many of the problems met with in old age cannot be understood in the light of old age itself, but that they have, to a certain extent, been created in the so-called productive years - or earlier.

If frequent family contact and a rare feeling of loneliness are considered desirable, one must ask how the relative contact-"privileges" enjoyed by "larger self-employed persons" can spread to other social classes, including, not least of all, "workers". Here it should be borne in mind that the "larger self-employed persons" have, generally, better economic and housing conditions than "workers", and the hypothesis can be put forward that the degree of isolation and loneliness which elderly "workers" experience would be reduced if it could be guaranteed that socio-political measures for elderly people would benefit, in particular, those who are under-privileged socio-economically, so that, the distribution of welfare on the whole would become more equal. It was found, for example, that "larger self-employed persons" have stayed overnight with their children more frequently than "workers", and it is to be assumed that this can be explained, among other things, by the former group's relatively larger houses or apartments, which make them better able to have family members remain with them for some days. But a more equal distribution of welfare will require, among other things, an egalitarian socio-political state intervention, and it is a question of whether possibilities of this exist at all.

Retirement, too, proved to have some adverse effects. It does not involve more frequent family contacts; on the contrary, it increases the experience of isolation and loneliness. A lowering of the pension age can, therefore, hardly be argued for on the grounds of the desire of a lower degree of isolation and loneliness among the elderly. It would probably be more reasonable, for instance from an activity and integration point of view, to point to a gradual withdrawal from the labour force, i.e., reduced working hours or easier/other work, adapted to the elderly person's own needs and wishes. But at present, it is in fact the "larger self-employed persons" who can remain in their jobs and benefit from shorter work hours, whereas "workers" most frequently have to adjust to the labour market by retiring from regular employment - thereby facing, among other things, increased isolation and loneliness. Thus, the effect of retirement on contact, isolation, and loneliness also contains elements of social inequality specific to class.

ABSTRACT

In this paper some results from the Danish longitudinal study of the elderly are presented and discussed. The results concern family contacts, isolation and loneliness in the early stages of old age, i.e., from the age of 62 to 73. These topics are discussed with special reference to social class. It is shown that the frequency of family contacts as well as the experience of isolation and loneli-

ness cannot be viewed independently of the social class affiliation of the interviewed persons. For example, it was found that "larger self-employed persons" tend to retain and continue to develop family contacts more easily than "workers". The paper also deals with the effect of retirement on the family contacts of elderly people.

REFERENCES

Olsen, H., Alderdom i klassesamfundet, *Socialt Tidsskrift*, Copenhagen, 1974, no. 12.

Olsen, H., Noter til en foreløbig bestemmelse af begrebet klassespecifik socio-økonomisk ulighed, *Socialt Tidsskrift*, Copenhagen, 1975, no. 8-9.

Olsen, H., et al., *Familiekontakter i den tidlige alderdom*, (Family contacts in the Early Stages of Old Age), Report no. 74, Danish Institute of Social Research, with an English summary and translation of figures and tables, Copenhagen, 1976.

Olsen, H., et al., *Ældres arbejdsophør* (Retirement from Work), Report no. 79, Danish Institute of Social Research, with an English summary and translation of figures and tables, Copenhagen, 1977.

Riley, M.W., A. Foner, *Ageing and Society*, vol. I: An inventory of research findings, New York, 1968.

Streib, G.F., C.J. Schneider, *Retirement in American Society - Impact and Process*, London, 1971.

9. Social needs of the relatives of old people

D. Schlettwein-Gsell and B. Bass (*)

INTRODUCTION

Problems of the older generation are usually discussed without considering the problems of the next-following generation although several publications make it seem probable that quite a few problems of the old generation could be solved if the situation of the younger relatives would be taken into account more carefully.

From Woodford-Williams (1965) we learned more than ten years ago the value of a relatives' clinic within a geriatric department. The mere fact that she had started to care for the problems of her patients' relatives enabled her to discharge up to 50 per cent of her patients to their own homes.

Richter (1970) made us realise that furthermore the solution of certain problems in the younger generation is crucial for the development of a successful partnership, which is not possible as long as the old man serves as an excuse for the young mans own incapability or as a target of his agressions.

In the last year several old age centers have started to analyse family involvement in nursing homes (York and Calsyn, 1977) or the role of the relative for the care of the aged person in a hospital setting (Rothstein, 1977). In the traditional Swiss centers these aspects, however, are rather uncommun up to this day.

(*) Foundation for experimental Gerontology, Basel.

It also appears that no study so far has dealt with the situation of the relatives of aged persons living independently.

We therefore considered it worthwhile to find out to what degree the problems of the relatives' generation were realised and what kind of support seemed practicable.

MATERIAL AND METHOD

One hour interviews were conducted with 100 female leaders of old age gymnastic groups all of whom were or had been younger relatives of aged persons themselves. We thus contacted a homogenous group of women aged 30 to 65 who besides their personal experience within their families had been exposed to old age problems by their occupation.

RESULTS

We list some of the questions and the respective answers:

Questions:	Answers (n = 100) yes	no	no answer
1. Is the younger relative of an old person frequently confronted with difficult problems?	86	8	6
2. Would it help the old persons if one cared for the younger relatives' problems?	94	5	1
3. Is the younger relative sufficiently informed on services offered to the aged?			
e.g. meals on wheels	19	53	28
day hospital	10	71	19
hospitalisation during holiday	8	81	11
ergotherapy	9	78	13
4. Which support seems practicable for the younger relative?			
individual advice	85	4	11
group-work	87	6	7
role-playing	65	16	19

	yes	no	no answer
5. Is it desirable to discuss more openly that the relation between the generations is shadowed by the fact that the younger relative might			
be afraid of getting old himself	45	20	35
feel guilty towards the older generation	66	9	25
6. Would it help to know more about the influence of the surroundings on the old persons physical state?	63	13	24

DISCUSSION

More than 90 per cent of the subjects agreed, that the younger relative is frequently confronted with very difficult problems by the mere existence of an aged person within his family. 94 per cent hoped for possibilities to support the younger relative.

The *information* on services offered to the aged by the community is rather incomplete even among this group of subjects specially interested in old age problems. 50 to 80 per cent stated that they were not informed about the services mentioned, which all provide considerable support also to the younger relatives of the aged. Blume (1972) had noted a similar disinterest among relatives of impaired children in Köln and claimed that the most disenable cared least for information but rather hid in anonymity.

Individual advice (85 per cent) and group-work (87 per cent) were accepted as *practicable for support* of the younger relative as is already practised with the relatives of psychiatric patients or parents of children with cancer. Role playing (e.g. answers to sterotypes like "you only wait for me to die" etc.) were considered helpful by 65 per cent of the subjects, the rest of them doubted the effectiveness of such methods in a Swiss population.

Two thirds of the subjects wanted increased information on *guilt feelings* of the younger relative and stressed the impact of such feelings on the well-being of the old person. It would seem important to analyse whether the kind of education (permissive or autoritarian) is related to the existence of such feelings in later life. *Being afraid of ones own age* was rejected aggressively (20 per cent) or refused to be taken into consideration (35 per cent) by the majority of the subjects. It seems necessary to analyse carefully whether existing problems might be hidden by such a rejection.

At the end of the interview we gave an example of the influence of the younger generations' attitude on the physical state of the aged. We told them that even incontinence can result from a desire for more tactile sensations and can be cured by more intensive and more careful tactile contact (Falck, 1972). 63 per cent of the subjects agreed that the information on such interrelations might considerably improve the younger generation's attitudes.

SUMMARY

Interviews were conducted with 100 female leaders of old age gymnastic groups. The impact of the aged on their younger relatives is felt by more than 90 per cent on this group. 50 to 80 per cent thought that more information on existing services was needed.

Individual advice and group-work seemed desirable as a support for the younger generation to more than 85 per cent of the subjects and might as a consequence considerably improve the situation of the old generation, as could be concluded from questions on guilt feeling, anxiety and general understanding of the aged person's situation.

REFERENCES

Blume, O., *Situation von Familien mit behinderten Kindern*, Forschungsinstitut für Sozialpolitik, Köln, 1972.

Falck, I., *Die Familie des alten Patienten*, Documenta Geigy, Basel, 1972.

Richter, H.E., *Patient Familie,* (Rowhohlt, Reinbeck, 1972).

Rothstein, U., The role of relatives in the care of aged persons in a hospital setting, Ystad Colloquium of the European Committee on social Gerontology, 1977.

Woodford-Williams, E., The value of a Relatives' clinic in the Geriatric Unit of a Hospital, in Busse, E.W. and Jeffers, F.C. (eds): Duke University Council of Gerontology, Proceedings of Seminars 1961-1965 (Duke University, Durham N.C. 1965), 286-297.

York, L.J. and Calsy, R.J., Family Involvement in Nursing, *The Gerontologist*, 17, 500-505.

Relations outside the family

10. Without or with a family

Monique Asiel (*)

INTRODUCTION

A previous survey, aimed at finding out the reasons why some people were capable of living independently despite their poverty, frailty and loneliness whereas others were ready to give up their way of life even though their health and financial status were good and their family present, permit us to consider other characteristics of an elderly female population (Asiel et al. 1975). The choice turns towards the influence of family relationships on individual behaviour. The study has been limited to a random sample of 480 subjects taken from the female population of Brussels, aged 70 and over, including only widowed, unmarried, divorced and separated women.

The first results indicate that the more one is able to rely on oneself, the easier it is to get along with others; the less one fusses about minor complaints and the more one is capable of admitting one's handicap, the more independent one is likely to be. But at the same time the study demonstrates that almost all the persons investigated live alone while maintaining good relations with their families. The members of the younger generation help their mothers or relatives when necessary. Nevertheless, 20 per cent of the subjects felt isolated due to the lack of a family or failure to maintain contacts.

(*) In collaboration with J. Decrucq, D. Florent, M. Lannoy and J. Pringels, Université Libre de Bruxelles, Faculté de Médecine et de Pharmacie, Laboratoire d'Epidémiologie et de Médecine Sociale.

METHODOLOGY

As stated previously, the random sample comprised the feminine population lists of Brussels city. Besides the criteria already mentioned, the subjects had to live in the heart of Brussels. A first list of 520 names was drawn up, but 14 names were those of deceased persons, 21 were women who had left the area, and 5 were new pensioners just admitted into an old people's home. Of the remaining 480 which represent the sub sample: 49 (10.2 per cent) had unknown adresses; 112 (23.3 per cent) were absent during the period of the interview; 5 (1.0 per cent) were hospitalized; 35 (7.3 per cent) refused the interview and 279 (58.1 per cent) accepted it.

The interviews were performed by three social workers from the staff of the "Laboratoire d'Epidémiologie et de Médecine Sociale" (Prof. Graffar) of Brussels University between the end of April 1975 and September 15th of the same year.

DESCRIPTION OF THE SUBSAMPLE AND REPRESENTATIVITY

On account of the important proportion of non-responses it is impossible to assess the validity of the sample. Still some criteria serve as good comparison points.

a. Age

The percentages according to age are similar in the sample and in the Brussels population. Two estimations are given to compare the data of the present survey.

Table 1. Age distribution

	Age groups						Total
	70-74	75-79	80-84	85-89	90 +	Unknown	
Subsample	92	83	59	34	9	2	279
	33.-	29.7	21.1	12.2	3.3	0.7	100.-
Brussels 1965	40.9	30.6	17.8	7.8	2.7	-	100.-
Percentages estimated 1976	36.8	30.4	20.2	9.7	2.9	-	100.-

Sources : Service de la Population Ville de Bruxelles
Asiel and Therer-Wollast, 1966.

b. Marital status

Regarding the marital status the situation is the same. The differences found with Dooghe's figures or those of the national census are due to the fact that this author's survey extends to the whole country. Accordingly, former estimated percentages for Brussels are very close to the present ones.

Table 2. Marital status among single women over 70 (in per cent)

	Marital status			Total
	Widowed	Unmarried	Divorced and separated	
Dooghe 1966	80	19	1	100
National Census 1970	83	15	2	100
Brussels 1965	72	20	8	100
Brussels 1975 (subsample)	75	17	8	100

Sources : Dooghe 1967, p. 12.
Annuaire Statistique de la Belgique, Tome 96, 1976, p. 33.

c. Social class

Since a long time Brussels is mostly an administrative and

commercial centre. Therefore, it is not surprising that there is a large percentage of women in trades and services.

Table 3. Occupation distribution (in per cent)

	Brussels 1975 (Subsample)
Housewives	29
Trades, Craftsmen	21
Independent	6
Employees	15
Workers	24
Home Helps	6

To avoid the imprecision of the housewife category, the professional status of the late husband or cohabitant of the subject was recorded. Employees as well as workers are then more numerous. This implies that a great number of the wives of the working class remain at home and might have fewer contacts with the outside world. Recent national statistics concerning retired women indicate proportions different from ours, but being national more than 6 or 7 per cent of the independents were farmers.

Table 4. Social class distribution (in per cent)

Social class	Social class	
	Brussels 1975 (Subsample)	Belgium Retired females
Liberal profession	2	-
Independent	28	40
Employee	27	11
Worker	43	49

Source : Annuaire Statistique de la Belgique, 1976, p. 521.

d. Education

What characterizes this population is its low eductional level. Indeed, 8 per cent of the interviewed are unable to read or to write. With respect to the age of the population this is normal since education is compulsory since 1914. However, this situation is changing. Ten years ago 80 per cent of the women of 65 and over did not attend school after the age of twelve. In our days this proportion decreases.

Table 5. Education distribution (in per cent)

Educational level	%
Illiterate	8
Primary school	61
Low secondary school	21
High secondary school	8
High school	1
Unknown	1

e. Financial status

Income in three quarters of the cases is considered sufficient. For the other quarter the financial status is poor: less than B.F. 8.000 per month. It is the share of former independents who were not obliged to contribute toward their future pension. Consequently, they receive the minimal pension from the State. Some of them are helped by the Public Assistance Commission or by their relatives.

f. Housing

Lack of comfort in the dwellings is another salient feature of the subsample. More than one half (56 per cent) live in an uncomfortable house or flat though 75 per cent of their owners or tenants have revenues which permit better housing. The buildings are rather old since 29 per cent of the inhabitants have not moved since 20 years or more. There is no hot water and most of them depend upon a coal stove for heat. The flats are usually located on the higher floors without a lift. Only 36 per cent of the group lives on the ground floor or on the first floor. In these cases, the maintenance is bad and dependent upon the low level of the rent. It is surprising that 46 per cent of the elderly women have a telephone at home: one third disposes of a telephone within the building, 6 per cent have asked for one or are on the waiting list, and the other 15 per cent are deprived of this mode of communication. In fact, telephones are found more frequently in the houses of the better educated women.

To summarize, the subsample is not very far from what one could expect in the Brussels female population. The group has a median age ranging between 75 and 79 years old. The great majority of those women are widows belonging mostly to the middle classes: employees and small tradespeople whose education level is low, though better than the general one of a group of former active Belgian women of the same age. The comparison is not so easy nor simple with respect to the last two criteria: pecuniary position and housing. Precise data are not available. What is known is what people are willing to express about the amount of pension they receive or the rent they pay. On the whole, the financial situation of the subsample is not bad even though it is not ideal especially when 25 per cent of the women have to manage with no more than B.F. 8.000 per month. Housing is not closely related to the financial situation of the dwellers. They seem content with mediocre lodging or dwelling as a result of at least twenty years of occupation. In order to investigate the influence of the familial environment on the behaviour of the elderly, a certain number of new criteria are taken into consideration.

1. Family

The majority of the elderly women interviewed have a family: 254 out of 279 or 87.4 per cent. But 36 women have discontinued contacts with their relatives though one of them replaces her family by friends. Likewise, of 25 old women without a family, mostly unmarried women, three have found a substitute family among their friends. Since, the subsample can be roughly divided into three categories of persons :

- those who have no family and no affective relations : 22
- those who have a family and no relations with it : 35
- those who have a family or a substitute family and positive relations :222

The latter is probably not very homogeneous but it seems to us that replaced families are just as warm as real ones.

2. Cohabitation

As two-generation household is not rare even in town. Out of the 222 women where a family or its substitute exists, 37 live with the family, 10 with friends. The remaining 171 live alone even when one of them perhaps shares her dwelling with a wage earner. This does not mean that 78.4 per cent of the group is abandoned since nearly 10 per cent have at least one or more relatives living in the same building. This coexistence is not bound to age. The same proportions of age groups are seen among those living together or in the same house or alone. No gradient is therefore demonstrated in cohabitation in the subsample.

3. Age

Neither is there an age difference between the group of 57 women who have no contacts with their family and neighbourhood and the 222 others just described.

Table 6. Age distribution and isolation

Age groups	Isolated	Non isolated
70-74	28	35
75-79	28	30
80-84	26	20
85-79	11	13
90 +	7	2
Total	100	100
N =	57	222

4. Mobility

Mobility decreases with age, but, as there is no difference in age between the isolated and the others, there is also no difference in their mobility.

Table 7. Mobility and age

Mobility	Age groups	
	70-79	80 +
Good mobility	59	42
Limited or walking stick	34	33
Need of help or chairbound	7	25
Total	100	100
N =	117	102

Chi square 17.33; $p < .01$

The importance of the handicap does not alter the attitude of the family towards the elderly with respect to cohabitation.

Table 8. Mobility and housing arrangement

Mobility	Housing arrangement		
	cohabitation or same house	living alone	without family
Good	42	54	46
Limited	41	31	42
Helped or chairbound	17	15	12
Total	100	100	100
N =	64	158	57

Chi square 3.25; $p > .10$

Even if there are fewer invalids among those who lack a family and the reverse among those who share the dwelling or the house with the family, the difference is not significative.

5. Health

Health is a factor which is almost impossible to define precisely in the subsample. Nevertheless, the interviews have not shown that isolated women are more inclined to use medicine. This is perhaps related to the frequency of consulting their doctor, since women without family relationships seem to consult the doctor more often than the others. If physical health is not very different in all the categories, the interviewer's impressions about mental health are more varied. Memory defects and behavioral troubles are more often encountered among women who have no family and especially among women who have discontinued relations with others. The difference is significant.

Table 9. Mental state and affective behaviour

Mental state	Affective behaviour		Total
	No relation	Positive relation	
Normal	69	85	83
Psychological troubles	31	15	17
Total	100	100	100
N =	35	222	257

Chi square 5.9; $p < .05$

Contrary to all expectations, sleeping habits and the use of soporifics are very similar in the two groups.

g. Behaviour of the elderly

If the presence of a family influences the behaviour of the elderly persons, how can the change of attitudes be tested? The aspect of the dwelling somehow reflects its resident. Whether the tenant or the owner was a lonely person the flat or the house was more often neglected than in the group of women with affective bonds.

Table 10. Aspects of the dwelling and isolation

Aspects of the dwelling	Isolation	
	Isolated	Non isolated
Very good + good	33	46
Mediocre bad + very bad	67	54
Total	100	100
N =	57	222

The tendency observed with respect to the habitations is really significative when applied to the personal appearance of the women. If 90 per cent of the women with a familial environment were well and even smartly dressed, only 74 per cent of the isolated ones were in the same state. It is not because of wealth that difference occured neither group was richer than the other. Nor was there a difference in the representation of the different social classes within the two groups; but, surprisingly, one observes that isolated women have received a better education. In this group 47 per cent have pursued their education after primary school, whereas in the other group only 27 per cent reached a higher level of education.

Table 11. Education and isolation

Educational level	Isolation	
	Isolated	Non isolated
Illiterate and primary	53	73
Higher education	47	27
Total	100	100
N =	57	222

Chi square 8.2; $p < .01$

It is not possible for the present time to assess a correlation between neglect and education; this requires further investigation. Isolated subjects devote their leisure time to more passive forms of recreation (T.V., radio, inactivity); others enjoy a more active life (walking, knitting, sewing). However, the difference is minimal. Free time is spent also on visiting friends, receiving them, attending third age clubs or ordinary associations. Here also the difference is too small to be significant. Both groups spend approximamtely the same amount of time on caring for an animal.

In conclusion, what are the reactions to the question about going to an old people's home. Not many persons accept this proposition; and if so, it is the more isolated who give an affirmative answer. But even then it is only a tendency.

Table 12. Opinion about entering Old People's home

Opinion about entering Old People' home	Isolation	
	Isolated	Non isolated
Yes	26	18
No	74	82
Total	100	100
N =	57	222

Chi square 3.4; $p > .05$

According to a previous work (Asiel et al. 1975) the motivations which inclined women to seek entry into an old people's home were: loneliness, lower level of education, medical abuses not always necessitated by a poor state of health, use of soporifics, fewer social contacts and a desire for group relations (club).

The conclusions of our present investigation are similar. Many factors influence the behaviour of people and the lack of a family can be a source of stress though not always. Some isolated women are as well adapted as women who enjoy a warm family milieu. Perhaps isolated people could or should be considered as a population at risk for certain needs. But even then, adaptation inherent to education, health, medical attitude, personality are present and play a role. Different characteristics exert different weights during one's life and the last one can be crucial; exterior influences have less power.

ABSTRACT

The aim of the present study was to gain a better insight into the influence of isolation or family contacts on the psychological behaviour of elderly persons. Therefore, a survey has been undertaken among a random sample of aged women living without a spouse in the city of Brussels. Different criteria have been examined according to isolation, real or deliberated. It appears that if the presence of a family can be a true and positive stimulus for the elderly women, there are also some isolated ones who are just as well adapted. This is probably due to the influence of various factors peculiar to the individual and tho his environment.

REFERENCES

Annuaire Statistique de la Belgique, Institut National de Statistique, tome 96, Brussels, 1976.

Asiel, M., J. Decrucq, D. Florent, M. Lannoy, J. Pringels, *Factors of Independence,* Communication Xth International Congress of Gerontology - Jerusalem, 1975.

Asiel, M., E. Therer-Wollast, Les besoins médicaux et sociaux de trois groupes déterminés de personnes âgées, *Proceedings 7th International Congress of Gerontology,* Vol. 6, 1966, 27-30.

Dooghe, G., *Les personnes âgées en Belgique,* Vol. I, Centre d'Etudes de la Population et de la Famille, Ministère de la Santé Publique et de la Famille, Bruxelles, 1967.

11. Loneliness of old widows and married women

Gilbert Dooghe and Lieve Vanderleyden (*)

A study in 1974, ordered by the National Council on the Aging, revealed that elderly persons are confronted with three important problems, namely, ailing health resulting in constraints of functioning and maintenance in society, insufficient economic resources, and loneliness. The following analysis focuses on the latter problem. An attempt is made to investigate how the loneliness phenomenon affects married people and elderly widows, and to explore the effect of some underlying factors. A path model was developed for this purpose.

THE DATA

The data used in this article have been extracted from a more comprehensive study on the life satisfaction of widows living alone and women living with a spouse only. For this study - the field work having been carried out at the end of 1976 - 240 female aged (one half of the sample widowed, one half married) have been interviewed. Institutionalized elderly individuals have not been included in this study.

LONELINESS AS A PROBLEM

As opposed to being alone, loneliness, is a subjective concept. It is a state of the mind, which is, of course,

(*) The Population and Family Study Centre, Ministry of Public Health and the Family, Brussels.

personally coloured. Loneliness occurs when an aged person experiences dissatisfaction as a result of his (her) actual social relations or the absence there of. This subjective experience is influenced by such factors as the objective fact of being alone, being childless or having children, psychological well-being, personal estimate of health, etc. Loneliness is strongly psychologically determined; it is a reaction affecting the mood and the feelings of the individual. Personal as well as social factors are important. The living conditions of the elderly often tend to make them suffer from loneliness. So, loneliness and old age are often associated. A number of reasons seem to indicate that loneliness increases with old age. Growing old goes along with the loss of relatives (death of one's spouse, brothers and sisters) and contemporaries (friends and acquaintances). The contacts and relations with children change when these are married and live independently, away from their parents. In many cases, old age implies a decrease of various somatic and mental abilities, preventing the aged from keeping up their social contacts. The marital status and especially widowhood has a determining impact on the mood of the individual and causes loneliness.

THE NUMBER OF LONELY PEOPLE

Determining whether the elderly were lonely was obtained through the analysis of the responses to the item : "I feel lonely". Table 1 indicates that loneliness is not a problem to every aged person. A significant number of widows living alone responded that they were not feeling lonely. This indicates that changes in the social and emotional relations of older people do not always result in loneliness. This does not alter the fact that, because of the special situation in which widows find themselves, it can be expected that those who live alone are more likely to suffer from loneliness than married women; this is evident from the research results in Table 1 and other studies (Shanas et al. 1968, 271).

Table 1. Loneliness among aged widows and married women (percentage distribution)

Marital status	"I feel lonely" Agree	Disagree	100 % =
Widowed	57	43	118
Married	13	87	120
Total	35	65	238

Somers's D = 0.43

Loneliness is more characteristic of widows living alone than of married women living with a spouse only. The likelihood that widows feel lonely is 43 % greater than with married people. The report on social reintegration of the aged in Austria (1976, 42) states: "In erster Linie ist das Vorhandensein oder Nicht-Vorhanden-sein eines (Ehe-) Partners ausschlaggebend für die subjective Beurteilung der Einsamkeit".

In agreement with the findings of some authors (Townsend, 1957, 175; Munnichs, 1964, 235) our study shows that there is a link between feelings of loneliness and the number of years which have elapsed since the death of the marriage partner. A significantly larger proportion of those who have been widowed less than five years (72 %), as compared with those widowed more than ten years (55 %) feel lonely. This indicates that for some widows the crisis caused by the loss of the partner is gradually declining whereas, for others, even after many years, widowhood marks the mood of the individual permanently.

The loss of beloved persons and loneliness seem to correlate strongly. Next to the recent loss of one's marriage partner, some authors (Townsend, 1957,173; Tunstall, 1965) indicate that elderly who have experienced the loss of one or more children, or who are separated from them, feel more lonely. Although the results of our study do not point in the same direction, it appears in any case that loneliness is more frequently characteristic of widows who have never had children. Of these, seven out of ten feel lonely; of widows with children the proportion is only 54 %. The low score of life satisfaction[1] indicates that loneliness reflects a situation of dissatisfaction.

Table 2. The average score of life satisfaction of elderly experiencing loneliness or not

Lonely	Widowed		Married	
	$\bar{x}$	N =	$\bar{x}$	N =
Yes	14.9	67	13.1	16
No	22.8	51	25.-	104
Total	18.3	118	23.4	120

LONELINESS AND BEING ALONE

Though being alone provides a favourable soil for loneliness, both concepts are not synonymous. Table 3 shows a marked correlation between the objective fact of being alone and loneliness, but does not indicate that all people living alone feel lonely. The likelihood of loneliness of widowed and married people who are often alone, is 28 %, respectively 35 % higher than for those who are not alone. Nor are all lonely aged people in an isolated situation. Persons who experience loneliness are 20 % more likely to be alone than those who do not feel lonely.

1. The Adam's Life Satisfaction Index A (1969, 470-474) was used to measure life satisfaction. For more complete data see: Dooghe, G., L. Vanderleyden, *Bejaarden en hun levensvoldoening*, 1978.

Table 3. Percentage distribution of aged people according to feelings of loneliness and being alone or not

1. Loneliness as a dependent variable

Feeling lonely	Widowed (*)		Married (**)	
	often alone	sometimes or never alone	often alone	sometimes or never alone
Yes	63	35	45	10
No	37	65	55	90
Total	100	100	100	100
N =	92	26	11	109

(*) Somers's D 0.28 $p < .02$

(**) Somers's D 0.35 $p < .02$

2. Being alone as a dependent variable

Being alone	Widowed (*)		Married (**)	
	lonely	not lonely	lonely	not lonely
Often	87	67	31	6
Sometimes or never	13	33	69	94
Total	100	100	100	100
N =	67	51	16	104

(*) Somers's D 0.20 $p < .02$

(**) Somers's D 0.25 $p < .02$

The fact that some isolated elderly do not feel lonely while some integrated aged do experience this feeling, is evidence by these concrete examples:

Case 45: The interviewed person is 73 years old and lost her husband 10 years ago. She has three married children and twelve grandchildren whom she sees regularly. She is very happy with the good relationship within the family. She experienced a lot of misfortune in her married life. Her husband had been a prisoner of war for five years. During this period, she worked as a maid to feed her family. Afterwards, her husband was often ill. Her family had always to live soberly, even poorly. Nevertheless, she was happy and found great consolation in her faith. When she becomes less mobile, she hopes to enter a home. She has no contacts with her brother or sister. Since they did not show up in difficult times, she does not want to see them now either. She says she is frequently alone, but never feels lonely.

Case 59: This refers to an approximately 86 year old married woman whose two children visit her weekly. She trusts her neighbour very much and her husband is very helpful. Although she is never alone, she nevertheless feels lonely.

Often a correlation emerges between the two separate phenomena of isolation and feelings of loneliness, as can be deduced from the next example:

Case 63: Mrs X is a widow, aged 70. She has cared for her husband during a period of seven years. This period has influenced her present situation: she is permanently depressed. She has lost her energy and constantly fears becoming dependent upon others. She is living in fear, has problems of insomnia and is under permanent medical control. She has very few contacts with friends or neighbours but sees her daughter daily. She says she is often alone and feels very lonely.

From the above cases, it becomes clear that the phenomenon of loneliness is very complex. Several factors and circumstances enter into play. In the following paragraphs, we will try to determine the factors that influence loneliness, and the effect of these factors on married women living with a spouse and on aged widows living alone.

MULTIVARIATE ANALYSIS

The aim of this article is to examine the factors which have an effect on feelings of loneliness and to determine the extent of the effect. Until now we have investigated the relation between being alone as an objective fact and loneliness as a subjective experience. To study the impact of a number of variables, a path model was developed that permitted detection of the intercorrelations of the variables included in the model.

Description of the model

In our model, loneliness is the dependent variable. The independent variables can be split into exogenous and endogenous ones. The exogenous variables are those variables which are not explained by the model itself and which have both a direct and an indirect effect on the dependent variable, i.e., loneliness. This does not necessarily mean there cannot be a relationship between those exogenous variables. This relationship, however, is not analyzed in the model. Exogenous variables are defined by factors not included in the model. The following three exogenous variables have been incorporated into the model:

- Age. It can be assumed that there is a relationship between age and loneliness. The incidence of loneliness increases with age among old people.
- Number of children. At advanced ages, having children can be considered as a protecting factor to isolation and loneliness. Numerous studies have indicated that the presence of children does influence greatly the social involvement of the elderly.
- Perceived financial adequacy. The way in which the financial situation is evaluated by the elderly, can have an impact on their morale and their behaviour. Unfavourable material conditions hamper integration into society.

Four endogenous variables have been taken into consideration. Endogenous variables are those by which the exogenous ones have an effect on the dependent variables.

- Initially, the degree of incapacity of the aged was to be taken into account. However, since the sample includes people living alone or with a spouse only, which implies that they can function on their own, the idea of using the physical fitness of elderly was dropped - measured on the basis of the degree of disability - as an endogenous variable. In order to include the health factor into the analysis, use was made of the variable "self-assessed health". Maddox and Douglass (1974, 56) are of the opinion that the perceived state of health is an important and useful variable in the study of the course of an individual's life, especially at the old age stage. Furthermore, they state that the perceived state of health can be more important than the objective medical diagnosis in determining the emotional behaviour of a person. A correlation between a negative health perception and loneliness can be assumed.[1]

- The degree of involvement in society was determined by the daily or less than daily contacts of the aged person with children, family members and non-relatives. Less than daily contacts with the outer world enhance the likelihood of loneliness.
- Another variable was the level of activity, measured on the basis of the number of hobbies. Low activity leads to monotony and boredom, and increases the likelihood of loneliness.
- As already mentioned earlier, being alone is a favourable soil for loneliness. Of course, this variable has been included in the path model.

The model has been applied separately for married women living with a spouse and for widows living alone. The effect of the marital status on loneliness has already been investigated. The question is, however, whether or not the influence of the aforementioned variables is the same with married people as with widows.

1. This analysis explores the effects of self-perceived health on loneliness. Although there might also be a relation in the opposite sense (loneliness and the feeling of not being involved any more can endanger the physical and mental health of a person), it is more logical to start from the relationship as state in the path model. The reciprocity of the relation cannot be denied but only the most evident direction of the relation is studied.

Table 4. Frequency distribution of the variables applied in the path analysis concerning loneliness (percentage distribution) (N = 240)

Variables	Frequency distribution
Age	65-69: 36 % / 70-74: 28 % / 75-79: 23 % / 80 and +: 13 %
Number of children	0: 20 % / 1: 34 % / 2: 23 % / 3 and +: 23 %
Perceived financial status	amply sufficient: 35 % / just sufficient: 50 % / not sufficient: 15 %
Personal estimate of health	very good: 19 % / fairly good: 59 % / bad: 22 %
Social contact score	daily: 59 % / not daily: 41 %
Number of hobbies	0 and 1: 8 % / 2: 19 % / 3: 25 % / 4: 26 % / 5 and +: 22 %
Being alone	often: 44 % / sometimes and never: 56 %
Feeling lonely	disagree: 65 % / agree: 35 %

With the exception of three variables, namely, the social contact score, being alone and loneliness - which have been dichotomized - all the other variables are measured on the ordinal level.

Bivariate analysis

The interdependencies of the variables in the model are reflected in the correlation matrix. This matrix shows the relationships among the eight variables and provides the basis for analyzing the causal model presented in figures 1 and 2. In the correlation matrix both positive and negative loadings are found. The highest correlation coefficient in the group of married people amounts to -.34 and +.34 respectively for the relation between the personal estimate of health and number of hobbies and for the relation between the personal estimate of health and loneliness. In the case of a more negative health rating, the number of hobbies decreases and the likelihood of loneliness increases. Among widows, one also observes that a negative health perception induces feelings of loneliness (r = .33). Of course, these correlation coefficients do not provide us with information concerning a direct link or an indirect one caused by other variables. In essence, the correlation coefficients represent the total effect between variables. To obtain an idea of the direct effect, a path analysis is performed.

The path analysis technique

This paper is an empirical study on the impact of a number of variables, which were built into a hypothetical model on feelings of loneliness. In no way are the results definitive, since the arrangement of the variables is incorporated in advance into the path structure by the investigator. As such, this arrangement has a hypothetical character. Use of other variables, or change in the structure of the path model would yield different results. The models shown in fig. 1 and 2 are recursive, which means that there are no reciprocal relations and that all correlations point in the same direction. To avoid complication of the model, only arrows with a path coefficient having a value of .10 or more, or being significant at .01 according to the F-test, are drawn. Both in the tables and in the figures, path coefficients and path regression coefficients are shown. The latter are reported in parentheses.

Path regression coefficients are unstandardized partial regression coefficients, which can be used if one seeks to compare cross populations, in our case married women and widows. Path coefficients, on the other hand, are standardized partial regression coefficients measuring the actual amount of impact that each variable has on the others in a given population. Through path analysis, one is able to decompose the total effect of the multiple variables included in the model, into a direct and an indirect effect.

Path analysis appears to offer excellent possibilities in bridging the gap between theory and empirical research. However, one must take care not to expect more than path analysis can deliver.

Table 5. Correlation matrix

Variables	X_1	X_2	X_3	X_4	X_5	X_6	X_7	X_8
Married (N = 120)								
X_1	1.00	-.12	-.07	.25	.03	-.18	.19	.04
X_2		1.00	.11	-.11	-.20	.04	-.14	.09
X_3			1.00	.15	-.03	-.12	-.04	-.03
X_4				1.00	-.09	-.34	-.13	.34
X_5					1.00	.05	.16	-.04
X_6						1.00	.05	-.30
X_7							1.00	-.30
X_8								1.00
Widowed (N = 118)								
X_1	1.00	-.12	-.02	.00	.06	-.10	-.12	-.15
X_2		1.00	.11	-.25	-.21	.18	.18	-.26
X_3			1.00	.07	-.10	-.14	.11	.11
X_4				1.00	.12	-.31	-.03	.33
X_5					1.00	.06	-.25	.16
X_6						1.00	.06	-.28
X_7							1.00	-.24
X_8								1.00

X_1: Age
X_2: Number of children
X_3: Perceived financial adequacy
X_4: Personal estimate of health
X_5: Social contact score
X_6: Number of hobbies
X_7: Being alone
X_8: Loneliness

Figure 1: Path model of some factors influencing loneliness of married women living with spouse

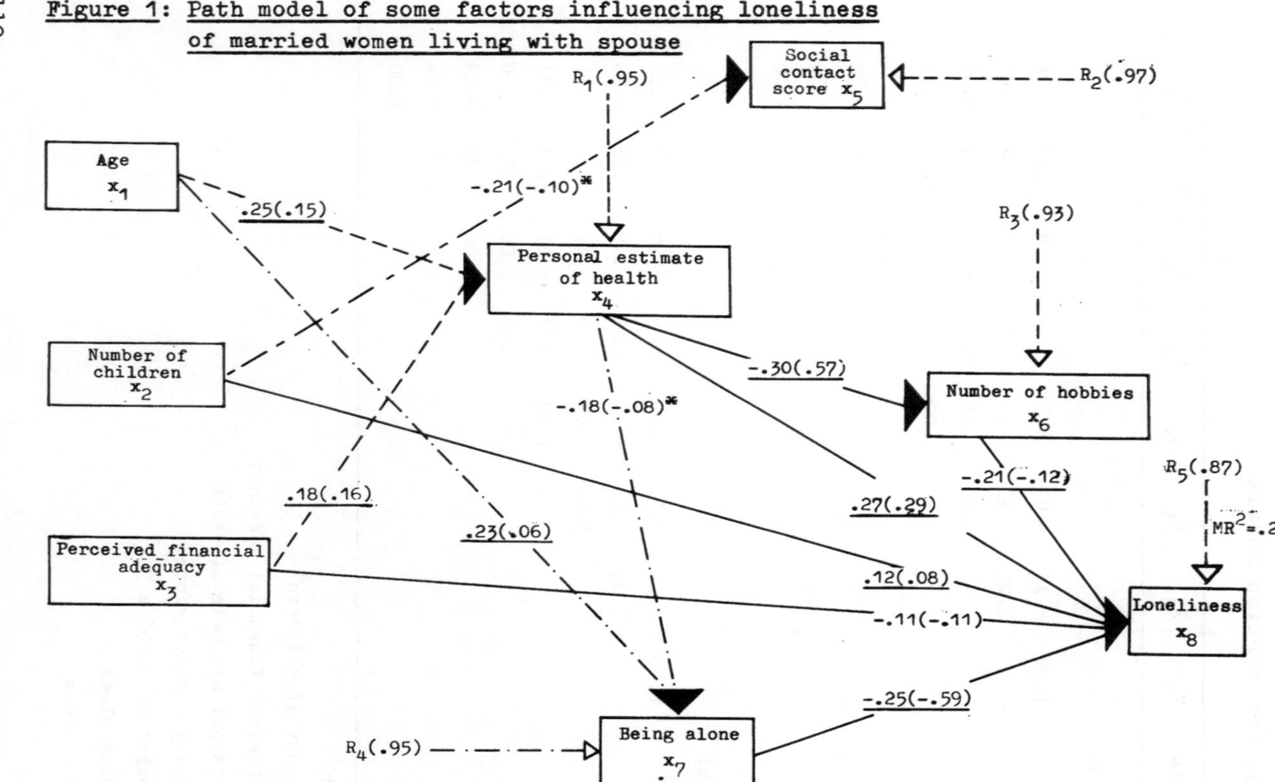

The first rate is the path coefficient, the rate reported in parentheses indicates the path regression coefficient. The values which are underlined are significant at the .001 level; asterisks mark coefficients that are significant at the .01 level; for the other values $.05 < p < .10$

Figure 2 : Path model of some factors influencing loneliness of widows living alone

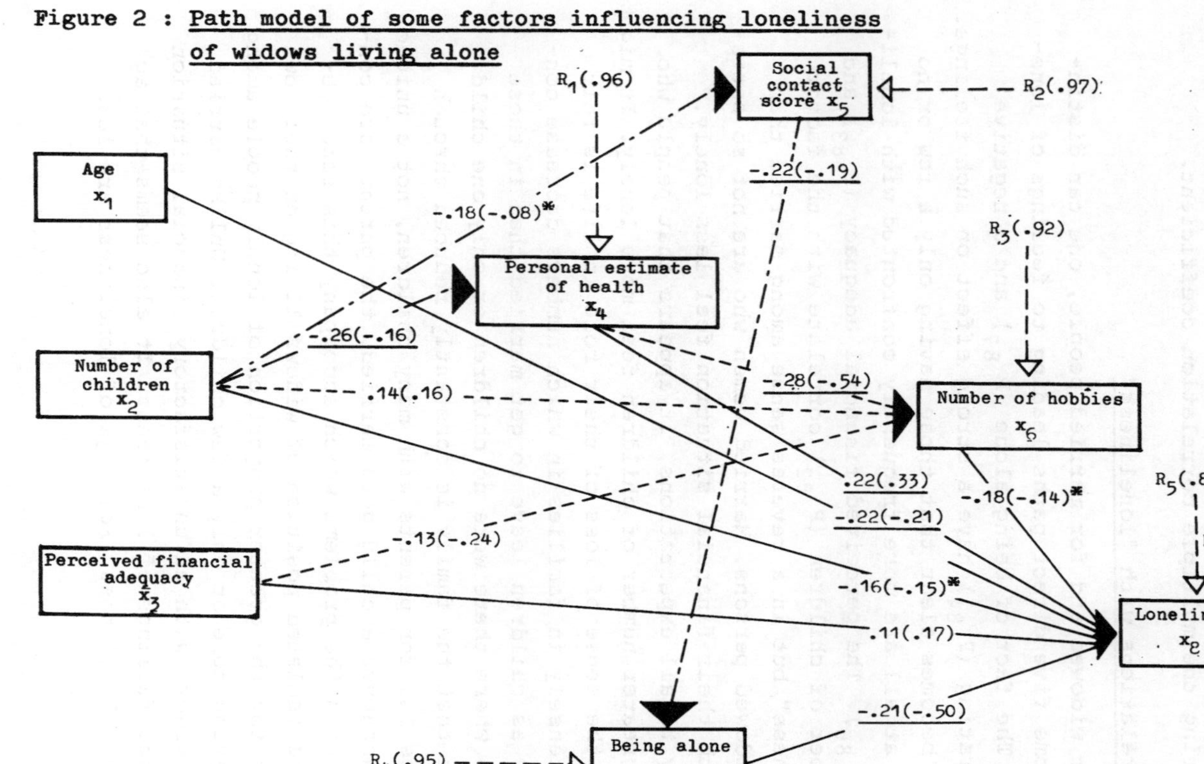

The first rate is the path coefficient, the rate reported in parentheses indicates the path regression coefficient. The values which are underlined are significant at the .001 level; asterisks mark coefficients that are significant at the .01 level; for the other values $.05 < p < .10$

RESULTS

The seven variables investigated for correlation with feelings of loneliness account for 24 % of the variation among married people ($F = 5.05$, $df = 7/120$, $p < .001$) and 27 % of the variation among widows ($F = 5.89$, $df = 7/118$, $p < .001$). The percentage of explained variance is obtained by squaring the multiple correlation coefficient.

Direct relations with "loneliness"

Both for widowed and for married people, one can distinguish some five direct paths leading to feelings of loneliness. The fact of being alone (P_{87}) and a negative health rating (P_{84}) have a strong effect on such feelings. Also it becomes clear that those having only a few or no hobbies at all are more frequently confronted with loneliness (P_{86}). The perceived financial adequacy (P_{83}) and the number of children (P_{82}) correlate with the factor "loneliness", but in a reverse sense among married than among widowed persons. Married women who are not so satisfied with their financial situation feel less lonely. Contrary to all expectations, it appears that people who have a greater number of children feel more lonely. It might be that the sense of loss of their role in life is felt more intensely in families in which numbers decrease continually as children leave to get married than in those families where there were no children or only one child. The fact that the family is constantly reduced through marriages is, for parents with many children, not a unique event. Whenever a child gets married, the parents are confronted with the problems which accompany this renouncement. The isolated position of widows living on their own accounts for the greater proportion of lonely people among those having none or only a few children. Unlike married people, widows with an unsatisfactory financial situation experience feelings of loneliness. It also seems that age influences loneliness. Feelings of loneliness are less

Table 6. Effect parameters applied in the path analysis concerning loneliness of married women

Variables	Personal estimate of health				Social contact score				Number of hobbies				Being alone				Loneliness			
	r	p	i.e.	p.r.c.	r	p	i.e.	p.r.c.	r	p	i.e.	p.r.c.	r	p	i.e.	p.r.c.	r	p	i.e.	p.r.c.
Age	.251	.251	.000	.148	.025	.031	-.006	.015	-.184	-114	-.070	-.130	.193	.229	-.036	.062	.043	-.011	.054	-.007
Number of children	-.108	-.098	-.010	-.059	-.195	-.205	.010	-.098	.045	.014	.031	.016	-.141	-.116	-.025	-.032	.090	.115	-.025	.076
Perceived financial adequacy	.150	.177	-.027	.160	-.034	.008	-.042	.006	-.119	-.083	-.036	-.144	-.039	.023	-.062	.010	-.026	-.113	.087	-.111
Personal estimate of health					-.088	-.120	.032	-.095	-.341	-.296	-.045	-.571	-.126	-.178	.052	-.082	.337	.269	.068	.293
Social contact score									.050	.026	.024	.064	.156	.111	.045	.065	-.040	.053	-.093	.072
Number of hobbies													.051	.035	.016	.008	-.296	-.214	-.082	-.121
Being alone																	-.300	-.250	-.050	-.588
Multiple correlation coefficient	.32				.23				.37				.32				.49			

r = product-moment correlation coefficient
p = path coefficient
i.e. = total indirect effect
p.r.c. = path regression coefficient

Table 7. Effect parameters applied in the path analysis concerning loneliness of widows

Variables	Personal estimate of health				Social contact score				Number of hobbies				Being alone				Loneliness			
	r	p	i.e.	p.r.c.	r	p	i.e.	p.r.c.	r	p	i.e.	p.r.c.	r	p	i.e.	p.r.c.	r	p	i.e.	p.r.c.
Age	-.001	-.031	.030	-.020	.065	.041	.023	.019	-.099	-.092	-.007	-.113	-.121	-.084	-.037	-.034	-.153	-.219	.066	-.212
Number of children	-.247	-.261	.014	-.159	-.215	-.180	-.035	-.082	.178	.135	.043	.159	.181	.112	.069	.043	-.263	-.163	-.100	-.151
Perceived financial adequacy	.071	.099	-.028	.099	-.105	-.090	-.015	-.067	-.141	-.126	-.015	-.243	.111	.081	.030	.052	.108	.112	-.004	.171
Personal estimate of health					.120	.082	.038	.061	-.311	-.282	-.029	-.543	-.028	.043	-.071	.027	.335	.216	.119	.330
Social contact score									.056	.112	-.056	.288	-.249	-.220	-.029	-.188	.157	.079	.078	.162
Number of hobbies													.063	.072	-.009	.024	-.276	-.177	-.099	-.140
Being alone																	-.238	-.210	.028	-.503
Multiple correlation coefficient	.27				.25				.38				.31				.52			

r = product-moment correlation coefficient
p = path coefficient
i.e. = total indirect effect
p.r.c. = path regression coefficient

frequently encountered among older widows than among younger ones. One could interpret this as an indication that widows, years after the death of the partner, have adapted themselves to this new situation, and have overcome the loss. For married women, the effect due to age is quasi nil.

Direct relation with other endogenous variables

The factors which influence feelings of loneliness are determined in turn, by a series of factors. It appears, e.g., that as they grow older married persons have a more negative health perception (P_{41}). This is not valid for widows: with advancing age they are not more likely to rate their health as less positive. Furthermore, it is clear that financial dissatisfaction (P_{43}) also has a negative impact on health perception, more so for married women (prc = .16) than for widows (prc = .10). While among married people age and perceived financial adequacy strongly influence the health rating, among widows the factor of having children or not is determining. The more children a widow has, the better the health perception.

Both for married and widowed persons, contacts with the outer world (P_{52}) are less frequent when they have none or only a few children; this indicates that the contacts and relations of the aged are influenced for a major part by the children.

Most remarkable is the link between the health perception and the number of hobbies (P_{64}). With married people and widows the number of hobbies decreases in relation to a negative health perception. Although not significant at the .05 level, widows with more children seem to have a greater number of hobbies.[1] Among widows, one observes also that a low financial satisfaction does

1. The bivariate analysis shows that 41 % of the widows having many children have five or more hobbies; for widows with only one child, this proportion declines to half the number.

not stimulate the number of hobbies (P_{63}).[1] It is obvious that a negative perception of financial adequacy accompanies a low activity level. In cases of low resources proportionally more of one's financial income is used to meet primary needs than in higher income groups, as that less remains to satisfy the defined cultural needs.

The relationship between being alone and a low contact score (P_{75}) emerges very clearly for widows but not for married people. The presence of the marriage partner provides a good explanation. Among married women being alone is less frequent with age increase (P_{71}). Enjoying the presence of the partner at such an age is an influencing factor. The knowledge of still being together and the awareness of each others mutual needs in daily activities and even more in situations of stress, such as illness and incapacity make partners dependent upon each other even more than before. Married people who rate their health as not so good, admit their loneliness more frequently. For widows living alone, the effect of declining health on feeling alone is minimal. Whether this indicates a state of resignation is an open question.

CONCLUDING REMARKS

In this paper, a path model has been used to determine the respective influence of some seven factors on feelings of loneliness of aged people. Data have been collected from a survey of 240 aged women, half of them widows living on their own, and another half living together with the husband. Although there is an evident relationship between the subjective feelings of loneliness and the objective situation of being alone, it appears that loneliness is parti-

1. Based on a survey of 409 aged widows, Arling (1976, 67) concludes that good health and the availability of economic resources are the most important factors which facilitate participation in a number of activities. This conclusion is very near to the results of our own survey.

cularly determined by the perceived health of the respondents. The path analysis has led to the following conclusions:

- The most important factors influencing loneliness are self-assessed health and being alone. The impact of those two factors seems to be greater with married people than with widows.
- There also seems to be a relationship between feelings of loneliness and the degree of inactivity. Elderly persons with few hobbies more frequently feel lonely.
- Among married people the sense of loss of one's role in life due to the departure of the children, has a strong impact; among widows the fact that they have none or only a few children is an important cause of loneliness.
- On the contrary, the likelihood of loneliness among widows does not increase with age. Most possibly, the loss of the partner is a moment of crisis. Studies in this connection have indicated that the death of the partner does undermine the resistance of the survivor, particularly during the first months (Young et al. 1963). Afterwards, one tries to overcome the loss of the partners and to adapt himself to the new situation.
- Furthermore, it appears that widows, to some extent dissatisfied with their income, indicate a larger proportion of lonely people. Obviously this is not the case for married women.
- The seven factors studied explain 24 % of the variance in loneliness for married persons and 27 % for widowed persons.

Although age is a very serious problem to people in general, it becomes even greater when maintenance and integration into society are endangered. Fewer contacts, a negative health perception - mostly the expression of a limitation of the range of action and of a less positive health status - and a decreased activity level giving rise to boredom lead to social stress, a very important factor in the early stage of loneliness.

REFERENCES

Adams, D.L., Analysis of a Life Satisfaction Index, *Journal of Gerontology*, 1969, vol. 24, no. 4, 470-474.

Arling, G., Resistance to isolation among elderly widows, *International Journal of Aging and Human Development*, 1976, 7, 67-86.

Die gesellschaftliche Reintegration älterer Menschen in Osterreich, Veröffentlichung des Bundesministeriums für Wissenschaft und Forschung, Springer-Verlag, Wenen, New York, 1976.

Dooghe, G., L. Vanderleyden, *Bejaarden en hun levensvoldoening*, Een empirisch onderzoek bij weduwen en gehuwde vrouwen, De Sikkel, De Nederlandsche Boekhandel, Antwerpen, 1978.

Lopata, H., Loneliness : Forms and Components, *Social Problems*, 1969, 17, 248-262.

Maddox, G., E. Douglass, Self-assessment of health, *Normal Aging*, II, Reports from the Duke Longitudinal Studies, 1970-1973, Duke University Press, Durham, N.C., 1974, 55-63.

Munnichs, J.M.A., Loneliness, Isolation and Social Relations in Old Age, *Vita humana*, 7, 1964, 228-238.

Shanas, E., P. Townsend, D. Wedderburn, H. Friis, P. Milhøj, J. Stehouwer, *Old People in Three Industrial Societies*, Routledge and Kegan Paul, London, New York, 1968.

Sheldon, J.H., *The Social Medicine of Old Age*, Report of an inquiry in Wolverhampton, London, Oxford University Press, 1948.

Townsend, P., *The Family Life of Old People*, London, Routledge and Kegan Paul, 1957.

Tunstall, J., *Old and Alone: A sociological Study of Old People*, London, Routledge and Kegan Paul, 1965.

Young, M., B. Benjamin, C. Wallis, The Mortality of Widowers, *The Lancet*, 2, 1963, 454-456.

12. The primary relations in old age. Children, brothers/sisters, other relatives, friends and neighbours

Kees Knipscheer (*)

INTRODUCTION

The most important point of view for our research into the social relationships of old age has been the idea that research in this field can't be limited to the study of the parent-child relationship but that all possible primary relations have to be taken into account.
Primary relations in this context include relations with children, with brothers and sisters, with other family members, with friends and with neighbours.
The main reasons why I have taken this point of view are the following:

1. Recent criticism of the research in the field of the relations between parents and their adult children (Gibson, 1972; Croog et al. 1972; McKinlay, 1973);
2. Some attempts to distinguish categories of primary relations and their distinctive functions (Adams, 1967; Babchuck, 1965; Litwak et al. 1969);
3. The discussion about an age-homogenous orientation among the old. (Rose, 1965; Rosow, 1967; Neugarten, 1970; Riley, 1972).

These three considerations have led us to collect, in the course of interview with elderly people, data about the five categories of primary relations already mentioned, namely children, brothers and sisters, other relatives

(*) Gerontological Center, University of Nijmegen.

with whom there is regular contact, friends, and neighbours with whom one is rather familiar.

GROUP OF RESPONDENTS

The data, which our analysis refers to, have been collected during a relocationstudy, performed by the Gerontological Centre of Nijmegen. This circumstance has had important consequences for the composition of the group of respondents. We cannot speak of a representative sample.
In the research the following sub-groups were distinguished: the people who were taken into a home for the aged between august 1974 and august 1975 (n=70), the people who had applied for the waiting-list for a home for the aged during the same period (n=186), a group of people who were on the waiting-list for houses, adapted for the aged, in september (n=101) and a group of elderly people who were living independently and had no intention of moving (n=98). The last group of elderly people, still living independently without any intention of moving, was considered as control-group within the relocationstudy.

The composition of the group of respondents as to sex, marital status and age is given in table 1. As expected, the widows/widowers and the people aged 70-80 appear to be overrepresented in the group of respondents as compared to the groups of people of 65 years and older in Nijmegen.

Table 1. The group of respondents as to sex, marital status and age.

		Age categories					
		up to 64	65-69	70-74	75-79	80 and older	total
Men	married	.4	9.5	9.7	6.8	4.2	30.5
	widowers	-	1.1	2.9	3.7	2.2	9.9
	divorced	-	.7	-	-	-	.7
	unmarried	-	.9	.9	.4	.2	2.4
Women	married	-	2.9	3.5	1.8	.4	8.6
	widows	1.3	8.1	11.6	10.1	8.1	39.3
	divorced	-	.4	.4	.2	.2	1.3
	unmarried	.2	1.8	2.4	1.5	1.4	7.3
Subtotal	men	.4	12.1	13.4	11.0	6.6	43.5
	women	1.5	13.2	18.0	13.6	10.1	56.5
Total	%	2.0	25.3	31.4	24.6	16.7	100.0
	abs.	9	115	143	112	76	455

A COMPARISON OF RELATIONSHIPS, A DYAD-ANALYSIS

By means of some recent example we can illustrate how much the interpretation of the nature and the significance of the intergenerational relationships still differ.
Referring to a comparative study in six countries (Denmark, Poland, England, U.S.A., Yugoslavia and Israel) Shanas (1973) states that the network of kinship relationships forms the most important source of social and psychological support for the elderly, and the relation between parents and children stands out as the most important by far. She mainly based herself on data about the presence of children (at least one), the living of at least one child or kinship relation in the neighbourhood and about the contacts with relatives taking place at least once a week. In the same year however, Zena Smith Blau takes a very pessimistic view on this. It is her opinion that grown up children fulfil their (material) obligations dutifully as regards their parents and do maintain contact, but that on the other hand the living separately, the different fields of

interest and daily experiences cause an alienation, which is carefully hidden by both parties (Blau, 1973). Referring to Schneider (1968) Blau points out that this alienation is hidden because it is contrary to popular standards. Rosemary too shows a rather pessimistic view in 1975. There are several possibilities to get a better understanding of the nature and the significance of the relationship between elderly people and their children. An extensive and exclusive study can be made of this relationship. One can also try to obtain more insight in the distinctive character of this relationship by comparing them to other relationships.

In the research mentioned above, the second possibility was chosen. A number of data was collected from the respondents of their primary relations. Besides the facts of sex, age, marital status, living distance and frequency of visiting, four functional relational aspects are distinguished. The first aspect is a very general one and is operationalised in one-item. The three other functional relational aspects, namely the interrelational functionality, the instrumental functionality and the "function" of intimacy are operationalised in several items, as mentioned.

Relational aspects		Relational aspect items
Birthday visit	R_1.	R_{11}: Who of these persons visits you on your birthday
interrelational functionality	R_2.	R_{21}: With whom do you play cards etc. R_{22}: With whom do you go out for a day R_{23}: With whom do you go walking
instrumental functionality	R_3.	R_{31}: Who helps you in the household R_{32}: Who does shopping for you R_{33}: Who does small jobs for you
Intimacy	R_4.	R_{41}: With whom can you speak confidentially R_{42}: With whom do you speak about his problems R_{43}: With whom do you speak about your problems

While the respondent had a review of the primary relationships mentioned by him in front of him, he was asked: Who of these persons visits you on your birthday? etc. First we will try out a comparative analysis of the 5 relation categories. This means that we are going to compare the relationships, mentioned by the respondents, grouped according to separate relation categories. The total of the relations given by a certain respondent can be called his primary social network. This primary social network consists of dyad-relations between the respondent and the persons he considers to belong to his primary relationships. In this terminology our analysis refers to the set of dyad-relations present for all respondents together and which are distinguished in 5 groups according the relation categories mentioned (Felling, 1974).
First of all we will give a review of the presence of the different relation categories in the primary social network of the respondents. Table 2 shows the percentage of the respondents with children (at least one), with brothers/ sisters, with other relatives and the percentage naming friends and neighbours. There were no respondents who did not name anyone in the relation categories

Table 2. The percentage of respondents who do and do not mention persons in a certain relation category.

	one or more	none
- children	76.-%	24.-%
- brothers/sisters	82.2%	17.8%
- other relations	70.5%	29.5%
- friends	71.-%	29.-%
- neighbours	44.6%	55.4%

The fact that more than half of the respondents don't mention neighbour-relationships is probably due to the fact that in a number of cases neighbours were considered friends (the interview-procedure was such that the question as to friends proceded the one as to neighbours).

The scores on the 4 functional relational aspects

The first phase of the data assimilation aimed at a reduction of the scores on the separate relational aspect items to scores on relational aspects. The procedure was as follows. If a person in the network got a positive score on one of the three items concerning a certain relational aspect, this was understood as a functioning of this person on that relational aspect. The result of this assimilation was a score on everyone of the four functional relational aspects for every dyad. A summary review of this is given in table 3. By means of this table we can establish e.g. that 86 % of the children pay birthday visits and that 32 % of the children take part in the field of interrelational functionality.

Table 3. Percentages of positive scores, per relation category and for the total, on the functional relational aspects.

	children	brothers/ sisters	other relatives	friends	neigh- bours	total
R_1-birthday visit	86	48	69	65	63	67
R_2-interrel. funct.	32	13	19	35	16	24
R_3-instrum. funct.	30	5	12	12	25	16
R_4-intimacy	65	37	45	64	44	52
Total (abs.)	1174	1150	980	1037	480	4794

It is noticeable that of the brothers and sisters it is always the lowest percentage that has a positive score. It is clear that the children are superior (3 times the highest percentage), but the friends as regards to one aspect have a higher precentage than the children and once they are almost equal.

To get a closer understanding in the correlation of the scores on the relational aspects, several association measures can be used. The distribution on the scales however created sometimes problems. Therefore a patternlike reproduction of combinations of positive and negative scores on the functional relational aspects was chosen. In table 4 the percentage distribution on these patterns per relation category is included.

Table 4. Relative frequency distribution per relation category on the patterns of positive and negative scores on the relational aspects.

1→ pos. 0→ neg. + n. appl. R3	 R2	 R4	 R1	 child- ren	 brothers/ sisters	other rela- tives	 friends	 neigh bours	Total % 	 abs.
1	1	1	1	15	2	3	4	4	6	288
1	1	1	0	-	-	-	1	-	-	18
1	1	0	1	2	-	-	-	1	-	32
1	1	0	0	-	-	-	-	1	-	4
1	0	1	1	8	1	4	4	8	5	218
1	0	1	0	1	1	1	1	3	1	42
1	0	0	1	4	1	2	1	5	2	112
1	0	0	0	-	-	1	1	4	1	43
0	1	1	1	12	7	9	16	6	11	504
0	1	1	0	1	1	1	5	1	2	77
0	1	0	1	2	3	5	6	3	4	175
0	1	0	0	-	-	-	2	1	1	41
0	0	1	1	24	16	20	23	16	20	975
0	0	1	0	4	10	7	11	6	8	370
0	0	0	1	19	19	25	11	21	19	899
0	0	0	0	8	39	22	14	20	20	996
TOTAL			%	100	100	100	100	100	100	
			abs.	1147	1150	980	1037	480		4794

What was already apparent in table 3 is confirmed in table 4. It appears that of the brothers/sisters it is by far the

highest percentage that have no positive scores on any relational aspect (pattern: 0000) and of the children the lowest percentage. For all relationcategories, more than half have no positive score on both the instrumental and interrelational functionality. The total of the four bottom patterns is 55 %, 84 %, 74 %, 59 %, 65 % respectively and for the total it is 67 %.
Here too brothers and sisters have the highest percentage, the children the lowest. Furthermore rather high percentages occur, proportionally in the pattern that state a positive score for all except instrumental functionality (0111). Especially outstanding is the 16 % for the friends. This again points to the part of the friends in the field of interrelational functionality. This is in contrast with the neighbours who have a proportionally high percentage in the pattern in which the interrelational functionality is not included.
Finally, it can be firmly stated that only a very small percentage have a positive score on all four functional relational aspects. For all relationcategories together this is 6 percent. Although the children have by far the highest percentage on this point, we must admit that even this percentage is low. In absolute numbers this means that only 176 of the total of 346 parents (109 respondents have no children) have a positive score on the four relational aspects. It must be remarked here however that a number of respondents declared that the instrumental functionality did not apply to them as they were still able to do everything independently, namely 26·9 %. (Through further network analysis it became clear that there are 108 respondents of which one or more children have a positive score on the four relational aspects). By means of tables 3 and 4 we can determine provisionally that we must fully agree with the conclusion of many gerontological research, viz. that the children form the most important category of the primary relations for elderly people. It must be added

however that the friends form a very important second category, especially as regards the inter-relational functionality and the intimacy. The neighbours specially play an important part in the instrumental functionality. Moreover is remarkable the less important part played by brothers and sisters. The other relatives, among whom uncles, aunts, nieces and nephews and grandchildren play a part that can not be ignored, but they have a less specific part.

Some gerontological associations tested, a dyad-analysis

As mentioned above we will look at our data on the level of dyads. Each relationship between a respondent and one of the persons in his primary social network we will consider as a dyad. These dyad-relationships can be analysed as to some characteristics of the respondent side, as to the side of the partner in the relationship and as to some characteristics of the relationship itself. Each of these three points of view will be taken in the following.

In the gerontological literature there are some results concerning social relationships of old people that seem to be accepted by many researchers and theorists in this field. The detailed information we have in our data about the primary social network will allow us to test some of these ideas. It is for example often found that there will be some differentiation in sex and marital status.
In table 5 we present our data on this point.
All significant associations between sex of the respondent and the scores on the relational aspects point in the direction of more positive scores for men, except as regards friends who pay birthday visit. It is striking that for the male respondents the children more often score positively on three relational aspects.
In the first instance this seems difficult to interprete. Maybe we should however compare this with the association as to marital status. It appears namely that for the married respondents the children score positively more often than for the widows/widowers and divorced (except for

Table 5. The percentages of positive scores on the relational aspects, differentiated for sex and marital status of the respondents, specified per relation category, including the significance of the differences.

	SEX OF RESPONDENT			MARITAL STATUS OF RESPONDENT			
	men	women	sign.[1]	married	unmarried	widowed divorced	sign.[1]
Birthday visit							
- children	91	87	x	91	-	87	x
- brothers/sisters	47	52		47	68	50	xxx
- other relatives	70	70		74	62	68	x
- friends	64	70	x	68	71	66	
- neighbours	63	67		63	49	70	x
Interrel. functionality							
- children	44	35	xx	40	-	38	
- brothers/sisters	17	18		15	21	19	
- other relatives	25	22		25	15	24	
- friends	43	42		43	30	45	x
- neighbours	21	20		21	16	20	
Instrum. functionality							
- children	43	35	x	43	-	35	x
- brothers/sisters	6	8		3	32	6	xxx
- other relatives	17	14		14	20	16	
- friends	17	16		23	21	12	xxx
- neighbours	33	33		29	39	33	
Intimacy							
- children	69	68		73	-	65	xx
- brothers/sisters	43	36	x	41	53	35	xx
- other relatives	57	37	xxx	54	38	41	xxx
- friends	65	66		68	69	63	
- neighbours	44	46		47	49	43	

1. Significance of chi-square:
xxx: .01 $<$ sign.
xx: .05 $<$ sign. $\leq$ .01
x: .10 $<$ sign. $\leq$ .05

the interrelational functionality). As most of the male respondents are married and only a small part of the female respondents (see table 1) it is possible that the married status of these respondents results in both the differentiation in marital status as well as in sex. For unmarried respondents brothers and sisters have a positive score more frequently on 3 of the 4 relational aspects. The other relatives and the friends sometimes score positively more often for the married. Finally we must state that the widows/widowers on the interrelational functionality have more frequently a positive score "from" friends.

Based on above information we tend to conclude that the married respondents have the best of it as regards the functioning of the various relation categories on the relational aspects. This conclusion seems to be more or less confirmed when we take the association with age. In general the younger age categories of the respondents have more frequently positive scores.

Table 6 presents the differentiations on the partner-side of the relationships. We have to pay attention at two points. First it is to be noticed that it is always the women who get a higher percentage of positive scores on the relational aspects. For the children this occurs on three relational aspects (interrelational and instrumental functionality and intimacy), for friends on two (birthday visit and intimacy), for brothers and sisters and for other relatives both on one.
In many gerontological literature the importance of the daughters is emphatically stated and often confirmed by research results. (Riley, 1968; Dooghe 1970). Although our data rather point this way we think we may conclude that they don't support an exclusive position of the daughter. (When asked who of the children had the most importance, 44 % of the respondents named a son and 56 % named a daughter).
As to the differentiation for marital status of the dyad-partners we can establish that the unmarried persons score positively more often, at least of the children and the brothers and sisters. With regard to the children it is sometimes a case of unmarried children living in with

Table 6. The percentages of positive scores on the relational aspects, differentiated for sex and marital status of the dyad-partners, specified per relationcategory, including the significance of the differences.

	SEX OF DYADPARTNER			MARITAL STATUS OF DYADPARTNER			
	men	women	sign.[1]	married	unmarried	widowed divorced	sign.[1]
Birthday visit							
- children	89	89		89	93	90	
- brothers/sisters	47	53	x	48	64	50	xx
- other relatives	70	69		70	82	60	xx
- friends	64	70	x	69	65	65	
- neighbours	59	68		66	56	67	
Interrel. functionality							
- children	33	45	xxx	38	52	27	x
- brothers/sisters	15	19		16	28	18	x
- other relatives	26	22		24	26	20	
- friends	42	43		41	40	48	
- neighbours	17	22		17	23	32	x
Instrum. functionality							
- children	33	45	xxx	38	56	31	xx
- brothers/sisters	7	7		6	26	4	xxx
- other relatives	12	18	x	16	14	18	
- friends	14	17		15	24	15	
- neighbours	38	31		38	14	25	x
Intimacy							
- children	64	73	xx	68	66	79	
- brothers/sisters	36	41		41	48	36	
- other relatives	45	47		46	48	50	
- friends	61	68	x	65	62	69	
- neighbours	40	48		42	47	51	

1. See table 5 footnote.

their parents. With regard to the brothers and sisters the functioning of the unmarried persons stands out the more as they are only positive to a small degree on the relational aspects in comparison to the other relationcategories. In view of this fact and remembering that for unmarried respondents brothers and sisters have a positive score more frequently on 3 of the 4 relational aspects, we may conclude that the functioning of the relationships between older brothers and sisters is minimal, except when one of the partners is unmarried. The third point we have to look at is the association of the scores on the relational aspects with differences between respondents and the dyad-partner. As stated in the introduction we are very much interested in this study in the importance of age homogeneous relations for older people. Our data offer the possibilities to check this to a certain extent. In order to do this we have determined the age difference between the partners for every dyad. It is apparent that the age differences for the dyads with the children and with brothers and sisters are strongly determined by the nature of the relationcategories. For the other relatives there is a wide spreading. For the friends there is an obvious tendency towards age homogeneity. If we draw a provisionary line at no more than 15 years younger for age homogeneous relations, 67 % appears to be age homogenous for the friends. For the neighbours this percentage is somewhat lower, viz. 54 % (Riley, 1968). It is obvious that the neighbour relationships show less tendency to age homogeneity, as the composition of the population in the neighbourhood is to a great extent decisive.
The fact that most of the friends and neighbours included in the primary social networks are of the same age group, says relatively little about the importance of the age homogeneous relations. In order to answer this question based on the dyad-data we have examined whether there is an association between the magnitude of the age difference and the scoring whether or not positively on the relational aspects. This association was only examined for the friends

Table 7. Percentages positive scores on relational aspects, specified as to age differences between dyad-partners, for friends and neighbours separately.

	5 years older than resp.	4 years older - 5 years younger than resp.	6-15 years younger	16-25 years younger	26-35 years younger	36-45 years younger	45 years younger	sign.[1]
Birthday visit								
- friends	63	70	64	71	68	65	50	
- neighbours	62	66	78	61	78	69	65	
Interrel. funct.								
- friends	43	51	46	38	19	19	30	xxx
- neighbours	14	28	25	18	19	13	7	
Instrum. funct.								
- friends	13	19	15	13	15	23	38	
- neighbours	25	19	30	35	50	50	30	x
Intimacy								
- friends	53	68	68	61	72	59	70	
- neighbours	48	45	52	39	53	42	45	
Total number of								
- friends	83	343	270	145	83	41	10	
- neighbours	29	106	125	82	38	38	19	

1. See table 5 footnote.

and neighbours, as in our opinion the age differences in the other categories are highly determined by the nature of the kinshiprelations.

By means of table 7 we can establish that there is no question of sensational associations. The part played by the friends in the interrelational functionality is mostly found for the age homogeneous dyads, but also occurs for the neighbours. For the instrumental functionality the tendency is in the opposite direction and most for the neighbours. Age associated tendencies are less or not to be found for the birthday visits and the intimacy. These percentages of positive scores are most varying. The fact that the data point most clearly in the direction of the importance of the age homogenous relations for the interrelational functionality, can be regarded as a certain confirmation of the findings of Blau (1960) and Rosow (1967). They found that the available people in the same position was a very important condition of a reasonable degree of social integration. Although nothing is known in our data about this availability, there is a tendency that in the field of interrelational functionality the age homogeneous relations with friends and neighbours play the most positive part. A second difference between the dyad-partner and the respondent we can construct out of our data concerns the occupational status. In other research it became clear that a large degree of social mobility of the children sometimes has a negative influence on the relationship with the parents (Adams, 1970; Guillemard, 1974). By means of the occupational status of parents and children it became possible to establish the difference in occupational status per dyad of parent and child.

In order to find out whether a large social mobility has a negative influence on the nature of the parent-child relationship, table 8 shows the percentages of positive scores on the functional relational aspects for the differences in occupational status.

It is obvious that this table does not show a consistent association between the degree of social mobility of the children and their functioning on the relational

Table 8. Percentages positive scores on relational aspects, specified as to differences in occupational status between parents and children.

Differences in occupational status	% positive on				TOTAL NUMBER
	birthday visit	inter-relational functionality	instrumental functionality	intimacy	
(1) child at least 3 levels higher	87	38	38	76	302
(2) child at least 2 levels higher	93	37	41	67	125
(3) child 1 level higher	90	39	44	66	165
(4) child same level	90	42	34	66	130
(5) child 1 level lower	90	47	36	70	46
(6) child 2 levels lower	92	44	47	71	24
(7) child at least 3 levels lower	80	20	17	75	6
(9) occupation of father or child unknown					349
					1147

aspects. From our data we have to take the conclusion that the parent-child relationships are not really negatively influenced by the social mobility of the children.

PROVISIONAL CONCLUSIONS

It must be strongly emphasized that the results given in the foregoing analyses are of a provisional character. Provisional in the sense of not being transferable off-hand to the respondents in our research.
In order to accomplish this we must perform a second phase of analysis in which we take the respondent's primary social network as unity of analysis. This analysis has not as yet been rounded off. A report on the results of this analysis will be made later on.

We shall give a brief summary of the comparison of the relationcategories, in so far as this is possible based on the foregoing analysis.

When comparing the functioning of the distinguished relationcategories one must always take the following into account. The inventory of the persons, considered by a respondent to belong to his primary social network, began by asking after the names of the children (all living children had to be named) and of the living brothers and sisters. After this he was asked to name other relatives with whom he had regular contact, friends (or good acquaintances) and neighbours with whom he had regular contact. This procedure implied that on the one hand all children and brothers/sisters are included in the data-information and on the other hand for the other relationcategories only those people were included in the network, who fitted into the chosen specification of these categories, according to the respondents. This question-procedure probably explains the fact that such a small percentage of the brothers/sisters score positively on the functional relational aspects. However the fact of course remains that these small percentages are a clear indication that the "collateral" family relations are of only slight significance compared to the other relationcategories. Only for respondents who are unmarried or married without children, can the brothers and sisters compete with the other relationcategories. It appears that the children mostly get the highest percentages of positive scores on the functional relational aspects. It can however not be disguised that only 65 % of the children talks to parents intimately. The functioning of the other relatives can be placed between the functioning of children and the functioning of brothers and sisters. There are not outstanding specific functions for this category.
The friends and neighbours fulfil more specific functions. The friends form an important category for the interrelational functionality and the intimacy. There are some

indications that especially the age-homogeneous friends relations are of importance for the interrelational functionality. The functioning of the neighbours is more concentrated on the instrumental functionality.

Table 9. Percentages of positive scores on the functional relational aspects, specified for friends living more than 15 min. away, friends living within 15 min. walking distance and neighbours.

Relational aspect	friends living more than 15 min. walking distance away	friends living within 15 min. walking distance	neighbours
Birthday visit	63	72	65
Interrel. funct.	37	49	20
Instrum. funct.	10	23	33
Intimacy	65	65	45

It seems that to a certain extent the respondents make an actual distinction between the parts played by friends and neighbours. This becomes stronger when we divide the friends into two groups: friends within 15 min. walking distance and friends who live farther away (see table 9). The specific differences between the neighbours and the friends living within 15 min. walking distance, appear rather larger than smaller.

ABSTRACT

An inventory of the total social networks of 455 elderly persons makes it possible to analyze comparatively the five different relation categories included, children, brothers/sisters, other relatives, friends and neighbours. These comparisons are not only made on the basis of structural characteristics, but also in terms of relationshipcontents, e.g. instrumental functionality, interrelational functionality and intimacy. It appears that the children get the highest percentages on functioning as to these relational aspects. The brothers and sisters function especially for those older people who have no children. The function of neighbours and friends is much less related to not having children. Friends seem specialized in interrelational functionality and intimacy. Neighbours are more important in the field of help and service.

REFERENCES

Adams, B.N., Isolation, function and beyond: American kinship in the 1960's, *Journal of Marriage and the Family*, 1970, 575-597.

Babchuck, N., Primary friends and kin: a study of associations of middle class couples, *Social Forces*, 1965, 483-492.

Blau, Zena Smith, Old age in a changing society, New York, 1973.

Croog, S.H., A. Lipson, S. Levine, Help patterns in severe illness: the roles of kin network, non-family resources and institutions, *Journal of Marriage and the Family*, 34, 1972, 32-41.

Dooghe, G., De structuur van het gezin en de sociale relaties van bejaarden, Antwerpen, 1970.

Felling, A.J.A., Sociaal netwerk-analyse, Alphen a.d. Rijn, 1974.

Gibson, G., Kin family network: overheralded structure in past conceptualizations of family functioning, *Journal of Marriage and the Family*, 34, 1972, 1, 13-23.

Guillemard, A.M., R. Lenoir, Retraite et échange social; tentative d'explication des systèmes de relations sociales en situation de retraite, Paris, Centre d'étude des mouvements sociaux, 1974.

Lehr, U., *Psychologie des Alterns*, Heidelberg, 1972.

Litwak, E., I. Szelenyi, Primary group structure and their functions: kin, neighbours and friends, *American Sociological Review*. 34, 1969, 4, 465-481.

McKinlay, J.B., Social networks, lay consultation and help seeking behavior, *Social Forces* 51, 1973, 3, 275-292.

Neugarten, B.L., The old and the young in modern societies, *American Behavioral Scientist*, 14, 1970-71, 13-24.

Riley, M.W. e.o., *Aging and Society*, Vol. III, New York, 1972.

Riley, M.W., A. Foner, *Aging and Society*, Vol. I, New York, 1968.

Rose, A.M., W.A. Peterson, *Older people and their social world, the Subculture of the Aging*, Philadelphia, 1965.

Rosenmayr, L., The many fàces of the family (University of Vienna), paper presented at the 10th. Intern. Geront. Congres Jerusalem, 22-27 july 1975.

Rosow, I., *Social Integration of the aged*, New York, 1967.

Schneider, D., American kinship: a cultural account, Englewood Cliffs, N.Y.: Prentice Hall 1968.

Shanas, E., Family-kin networks and aging in cross-cultural perspective, *Journal of Marriage and the Family*, 35, 1973, 505-511.

Townsend, P., *The family life of old people*, an inquiry in East London, London, 1957.

13. Typology of need for community services under the aspect of civil status and family relations of the elderly

Margret Dieck (*)

This paper deals with the knowledge available on family life in old age, in so far as the facts known are to be taken into consideration in planning community services providing aid for the elderly themselves and for their families. It also aims at demonstrating the use an economist can make of sociological and psychological research in gerontology and may contribute to the process of defining fields of interest in which greater and more conclusive knowledge should be acquired.

Orderly planning of area community services depends on knowledge relative to the use of social services. Besides health, housing conditions, educational background, the former working situation and income, the civil status and family relations are important factors determining the conditions of life in old age and the need of organized community services. One reason for dealing with the subject is to try to find answers to the questions as to which services will have to be provided in order to meet future needs. We will accept the customary division between social services that provide institutional care and support and those which provide support for the functioning of older people in their own homes. At the same time, we will try to differentiate between services needed by those elderly who live entirely without support from family members or friends in their own homes and services upon which the family may draw if older members are to remain in their

(*) Deutsches Zentrum für Altersfragen, Berlin W. West Germany.

own homes or in the family setting, although they are in need of continual aid.

The development of the age structure of the population is one main factor determining the volume and types of community services needed. In West Germany, the proportion of the population, 65 years of age and over, is estimated to be at the same level in 1990 as in 1975. At the same time, the absolute number of the old will diminish owing to a general decrease in population. Nevertheless the structural changes are important. In the span of time under consideration the proportion of older women will increase considerably: for every 100 men of 65 years and over there are today 166 women of the same age group; by 1990 this figure will rise to 206. In addition, there will be a decrease in the number of the "young old" (between 65 and 75 years of age) and an increase of the "old old" (over 75 years) by 30 %. The highest rates of increase are expected in the age group between 85 and 90 (Deiniger, 1976). As a consequence of this change in population structure, there will be a greater number of old women living without a spouse, for whom all types of services will either have to be provided by adult children and friends or by community programs.

A great number of investigations in different countries has shown, that sickness and different forms of disability and physical impairment are accentuated beyond the mid-70's and particularly in the case of surviving older women. After the age of 75, an increasing number of old people are restricted in their physical mobility, the number of those who are home bound and bedridden rises signficantly (Susser, 1969; Altenhilfe, 1974).

Taking into account the future change in the structure of the old population and the fact, that already today the availability of beds in institutions for between 4 and 5 % of the aged is insufficient, either the number of beds will have to be increased considerably or forms of

community service will have to be developed, which will prolong independent living even in the case of increasing disability. We also have to take into account, that there is a general trend to keep the elderly in their own homes as long as possible. A Danish study of the needs of the aged by Svane (1972) has shown, that at an average about 8 % of the population, aged 70 and over, had an unmet need for home help and that the total need for home nursing can be estimated at about 2 % of the whole population 70 years old and over.
According to Shanas (1969), between 8 % and 15 % of the elderly population living at home (depending on the country under investigation) need a full spectrum of services. An additional 6 to 16 % of the elderly can only go outdoors with some difficulty. These figures give an idea of the demand for services to be expected and for which we are, at least in West Germany, wholly unprepared. Existing service networks, even today insufficient, will not be able to cope with these numbers.

When we examine the living arrangements of the older population, we see that already today approximately half of those aged 65 years and over are either widowed, divorced or single and of these half live independently in their own household; the other half are married and of these over 70 % live independently in their own household. In other words, of the 75 % of the older population with adult children, only about half live together with their children under the same roof. There is a growing trend for the elderly to keep up their own household even after the death of their spouse rather than move into one household with the adult children (Dieck, 1974; Wirtschaft und Statistik, 1976). Even when aged parents and adult children live in separate households, it frequently occurs that most older parents live within proximity of at least one child and that even under such living arrangements there exists a high level of interaction. Widespread exchanges of material support and reciprocity still commonly link older parents with their adult children (Blume, 1968; Rosenmayr, 1975; Riley et al. 1968). Nevertheless, increasing geographic mobility of at least the younger family members may raise barriers to the administration of frequent and extensive services to the elderly. We do not know, how large a burden services toward the aged may become, in order to still be tolerable for both generations of one family.

One important factor which diminishes the extent of help that can be given to older parents and probably also determines the type of help still tolerable is the prevalence of salaried employment of women. In West Germany, the percentage of women of all ages employed has risen to an average of 39. For women living single, the figure rises to 59 and for widowed or divorced women it sinks to 20, while married women are employed at the average level. More than 70 % of the women gainfully employed give financial reasons for seeking occupation outside the household (Wirtschaft und Statistik, 1976). It is estimated, that these figures will continue to rise in the following decades, thereby making the care of aged relatives more and more difficult and lowering the probability that such help can be given extensively. Isaacs draws attention to the fact, that the source of support of disabled old people is increasingly moving from daughters to spouses and other old people - a fact, that may be the result of this development. In addition, even after giving up salaried employment, women may not be ready to resume their former role as a household and family "all rounder", as Rosenmayr points out in referring to the development in Eastern European countries. (Isaacs, 1975; Rosenmayr, 1975).

When asked from whom they would receive help in case of need, older parents very frequently name their children. According to various research findings, female parents are more likely to receive help from their children than male parents. Older parents of both sexes will more often receive help from their daughters and daughters-in-law than from their sons and sons-in-law (Riley et al., 1968;

Rosenmayr, 1973). This fact may be expected to coincide positively with the predicted rise in the number of older women. But even then, with a growing proportion of women with gainful employment, a rise in demand for supportive services should be a fairly sure prediction.
We do not know, to what extent there is a discrepancy between the help that is expected by older parents from their children, especially with respect to nursing care, and the help actually given in cases of need. The high expectancy of help found does not totally correspond to the results of an investigation into the living conditions of housewives in West Germany, who have no salaried employment. After all this is the group, with the most favourable position for taking over such responsibilities. According to Pross, who conducted this representative study, only 10 per cent of the housewives interviewed admitted sharing part of their time in nursing the sick or looking after old people. If there is any nursing activity, it amounts to an average of 11 hours per week (Pross, 1976). Rosenmayr (1973) reports, that the Austrian microcensus revealed rather low help activities. Only 22 % of the population, aged 80 years and over, received help in the household approximately once a week. Help in shopping was given twice as often. Can it not be concluded, from these figures, that extensive supportive services are not very frequently given?

In two pilot studies, carried out in Köln and Berlin, Lange examined the living conditions of low income families who looked after and nursed older family members living under the same roof (Lange, 1973). Husbands are in general nursed by their wives, older females by their daughters, followed by sisters and friends. Grand-children seem to be involved only to a negligible degree, a fact, wich may seem astonishing if one considers that in a growing number of cases four generations of a family are alive and that adult grand-children could share in the support of their grand-parents. According to these two studies, about half of the family members nursing an old relation are occupied with this task between two to six hours per day. Every third person gives attention to this activity all day. Continued nursing over a length of time leads to disturbances in family life, reduces the frequency of contacts with non-household members and is a continual strain for the significant caring person. The relationship between the person cared for and the nurse do not seem to undergo as strong conflicts as is the case of the relationship between the nurse and her husband and children. Long-term caring for an older relative seems to affect family life negatively. After experiencing the load which long-term nursing and supportive services entail, many women state that in their own old age they will not depend on their own children for such extensive services. Although these facts are not in favour of the family network contributing mostly to the support of disabled older people, the possibility of taking care of older parents is still considered to be one major advantage of the multi-generational family. Adult children are, in many cases, willing to reestablish a common household with their parents after one parent has died and if the older family members are in poor health. Low income is a factor promoting such solutions. In general, the intensity of reciprocal support between older parents and their adult children seems to diminish with a rising income. And with the rising income, too, the living arrangement preferred by incapacitated older people is not the nursing home or living together with family members, but the desire to remain in their own home with nursing care provided (Riley et al, 1968; Blume, 1968). It may be asked, if in the first instance, whether financial reasons and the lack of acceptable alternatives do not force adult children to support older family members to the extent they do. This argument seems to be underscored by the other well-known fact, that many of the elderly with adult children who move to old age homes and nursing homes do so because they do not wish to be a burden to their children and their families and thereby strain family relations.

Considering the above mentioned change in the population structure and the increase in salaried employment of women, and taking into account that the family is even today the main source of support for the elderly, help for the aged should not focus on the aged living alone, but must also consider the support to be given to families. As Bernice Neugarten has put it recently, we have to think how to best maintain the family that is still the most important social institution of all in meeting the challenges of the age (Beverly, 1976).

Clearly, a great part of the elderly with friends or family members who are ready to provide help and aid, either under the same roof or within proximity, may be classified as a population, for whom institutionalization and "open care" are alternatives, on the condition that an extensive system of domiciliary and nursing aids can be built up. In order to determine which types of care are needed, depending on civil status and family relations, some data at least are available.

Of the population living in institutions for the aged, the vast majority (about 70 per cent) are women who formerly lived alone - widowed, divorced or single. Older married couples very rarely tend to move to an old age home or to a nursing home. Older men, who number low in absolute terms, are nevertheless over-represented, a fact, which may be explained by the more difficult role of the aging male, which Rosenmayr (1973) points out. Very clearly the "old old" living alone form the main part of the population living in institutions.

In comparison, the "young old" are to be found in "normal" living arrangements or in the different forms of shelter housing. The same is true for married couples. Household help and home-nursing-services as well as different types of food service will most probably be types of support, which will suffice in many cases to ensure independent living arrangements over a long span of time. Even women with salaried employment may be able to offer the amount of househelp and of shopping services needed by this category of the old. Friendly visiting programs and old people's clubs are types of services that may help to ameliorate the situation.

Comprehensive social services and health services, which are real alternatives to institutionalization, for those elderly who are no longer able to cope with the demands of keeping up an independent household, who are chronically ill and/or physically disabled and in need of constant nursing care, cannot be delivered on a broad basis. If there are family members, friends or neighbours willing and able to render such extensive services, they are in my opinion the only alternative solution to institutionalization. We must bear in mind, that a certain percentage of the elderly cannot resort to such forms of help. According to Blume, they amount to at least 4 percent of the older population (Blume, 1968). If institutionalization is not to be the solution in cases in which alternatives are available, it seems of greatest importance that the full load of supportive measures necessary, including nursing, is not to be imposed upon those willing to help, so that society may benefit from their goodwill for months and even years.

In such cases, the above mentioned types of service may be a certain help, but they will surely not suffice if such help must be given over a long stretch of time. Short in-residence hospital visits may at least enable the significant caring person to have a holiday. Visiting nurse programs may be a considerable help. Besides, day clinics and day centers are services which will enable the adult children, the spouse, or whoever renders the services to lead at least partly normal lives. Even salaried employment of the younger female members of the household can possibly be combined with extensive support under such conditions.

To my knowledge, no data on the civil status and the family relations of those elderly, who in the first instance make use of holiday beds in institutions, are available. But there are a few investigations into the social structure of patients visiting day care arrangements, which seem to show a clear difference between the population in institutions and the patients treated in day clinics. As Huber reports, about half of the patients treated at the geriatric day hospital of the Felix-Platter-Spital in Basel live together under the same roof with family or friends. The data does not give any information on supportive services given to the other half reported to be living alone (Huber, 1974). Bergmann compares the civil state and the living arrangements of in-patients and day-patients of a gerontopsychiatric service in Newcastle. Although no firm conclusions can be drawn on the strength of this data, slightly more married patients and more patients still living in their own home (and in institutions) are admitted to the day-hospital in comparison with in-patient hospital treatment (Bergmann, 1972). The figures reported by Brocklehurst in his investigation of five day-hospitals in Great Britain and Northern Ireland show that 71 % of the male patients and 28 % of the female patients are married. As to the living arrangements, only 26 % lived alone, 39 % lived together only with members of the same generation and 21 % with members of two or three generations. It may be supposed, that the high proportion of widows who visited the day-hospitals did for the major part not live alone. In comparing his findings with Townsend's survey of old people in Bethnal Green, Brocklehurst comes to the conclusion, that day-hospital patients are much more isolated from their families than other old people (Brocklehurst, 1973). We cannot fully agree to this, since even the scarce data listed clearly indicates differences between day-care patients and the elderly living in institutions, as far as family relations and civil status are concerned. Our conclusions are more in line with Rathbone-McCuan and Levenson, who studied in-patients and out-patients on the levels and patterns of role performance. This study reports more engagement of day-care participants in family-related social roles. It also states, that day-care participants had living arrangements which provided them with daily access to their families (Rathbone McCuan, et al. 1975).

If I have not overlooked important investigations in this field, the data we have in relation to the types of services needed for different groups of the older population, depending on their living arrangements and family background, is very scarce. The little information we have about the elderly and the use they make of services provided shows marked differences, depending on the type of need. The development to be foreseen can leave no doubt as to the importance of planning in this field, but no careful planning is possible, so long as there is still such a lack of sufficient knowledge.

REFERENCES

Bergmann, K., Psychogeriatric Care in Great Britain with special reference to the Place of the Day-Hospital, *Gerontopsychiatrie 2*, Datensammlung, Dokumentation, Klassifikation, Therapie, zweites Symposium der Arbeitsgemeinschaft für Gerontopsychiatrie in Zusammenarbeit mit der Sektion Gerontopsychiatrie der World Psychiatric Association, Berlin, 12/13 Mai 1972, Janasen Symposien Bd. 9, Düsseldorf.

Beverley, E.V., Confronting the Challenge of Dependency in Old Age, British and American Experts on Caring for the Elderly Meet to Share Their Experiences and Opinions, *Geriatrics* 31, 1976, 7, 112-119.

Blume, O., *Möglichkeiten und Grenzen der Altenhilfe*, Tübingen 1968.

Brocklehurst, J.C., *The Geriatric Day Hospital*, A Report of Three Studies of Geriatric Day Hospitals in Great Britain and Northern Ireland, published by King Edward's Hospital Fund for London 1970, 2nd impression 1973.

Deiniger, D., Die zukünftige Entwicklung des Anteils älterer Menschen an der Bevölkerung unter besonderer Berücksichtigung der Sozialhilfeempfänger, *Nachrichtendienst des Deutschen Vereins für öffentliche und private Fürsorge*, 56, 1976, 8, 212-217.

Dieck, M., Entwicklung der Privathaushalte nach Zahl und Struktur, Ergebnis der Mikrozensus aus der EG-Arbeitskräftestichprobe 1975, *Wirtschaft und Statistik*, 1976, 7, 424-428.

Dieck, M., *Wohnen und Wohnumfeld* - Bedingungen des Wohnens älterer Menschen in der Bundesrepublik, hrsg. von Institut für Altenwohnbau des Kuratoriums Deutsche Altershilfe e.V., vervielf. Manuskript, Köln, 1974.

Erwerbstätigkeit im Mai 1975, Endgültiges Ergebnis des Mirkozensus, *Wirtschaft und Statistik*, 4, 1976, 230-236.

Huber, F., Das geriatrische Tagesspital mit besonderer Berücksichtigung der Erfahrungen der Tagesklinik im Felix-Platter-Spital, Basel, *Aktuelle Gerontologie*, 4, 1974, 369-379.

Isaacs, B., Geriatric Medicine in the United Kingdom, *10th International Congress of Gerontology, Congress Abstracts*, Vol. I, Jerusalem, 22-27 June, 1975, 240f.

Lange, U., *Der Enfluss der Pflegebedürftigkeit chronisch kranker älterer Menschen auf die Familiensituation im Mehrgenerationshaushalt* - Eine sozialempirische Studie im Stadtgebiet Köln, hrsg. von Institut für Sozialforschung und Gesellschaftspolitik, vervielf. Manuskript, Köln, 1973.

Lange, U., *Pflegebedürftige alte Menschen in Berlin* - Probleme ihrer hauslichen Betreuung, hrsg. von Institut für Sozialforschung und Gesellschaftspolitik, vervielf. Manuskript, Köln, 1974.

Pross, H., *Die Wirklichkeit der Hausfrau*, Die erste repräsentative Untersuchung über nichterwerbstätige Ehefrauen: Wie leben sie? Wie denken sie? Wie sehen sie sich selbst?, rororo-Taschenbuch, Hamburg, 1976.

Rathbone-McCuan, E., J. Levenson, Impact of Socialization Therapy in a Geriatric Day-Care-Setting, *The Gerontologist* 15, 1975, 4, 338-342.

Riley, M.W., A. Foner, *Aging and Society*, Vol. I: An inventory of research findings, New York, 1968, 550f.

Rosenmayr, L., Family relations of the elderly, Recent data and some critical doubts, *Zeitschrift für Gerontologie*, 6, 1973, 4, 272-283.

Rosenmayr, L., The many faces of the family, A critical appraisal of functions for the elderly, *Paper delivered at the 10th International Congress of Gerontology*, Jerusalem 22.27, 1975.

Shanas, E., Measuring the Home Health Needs of the Aged in five countries, *8th International Congress of Gerontology*, Washington, 1969, Proceedings, vol. 1, 260-262.

Susser, M., Abschlussberichte der interdisziplinären Untersuchung den Gesundheitszustand älterer Menschen unter Berücksichtigung ihres sozialen Status und ihrer gesellschaftlichen Kommunikation, *Altenhilfe* 2, Bericht der Landesregierung, Hrsg. Ministerium für Arbeit, Gesundheit und Soziales des Landes Nordrhein-Westfalen, 1974, 49ff.

Susser, M., Aging and the field of publich health, in: Riley, M.W., et al. (Ed.): *Aging and Society*, Vol. II: Aging and the professions, New York, 1969, 114-160.

Svane, O., Determining the Needs of the Aged, *9th International Congress of Gerontology*, Kiev, July 2-7, 1972, Reports, Vol. 2, 204-207.

14. Family helping patterns in a local Swedish retirement-club

Lars Tornstam (*)

This report is based on a specific study of one single retirement club in Sweden, a local club of PRO, which is one of the two dominating retirement organizations in Sweden. PRO has about 300.000 members and is organized in 1.243 local clubs. It is one of these local clubs which has been the target of investigation in this study. From the beginning the aim was to make a comparison between retirement organizations in Sweden and Poland. Unfortunately, this comparison has not been carried out; only the Swedish part of the investigation has been carried through, but the Swedish material contains a great deal of interesting data, which gives it a value in itself.

The empirical investigation is based on a 50 per cent random sample from the members' file of the local retirement club, which contained 1.132 members. Due to the great number of dropouts (about 40 per cent) it is very difficult to make any safe and sound generalizations from the data base. The total number of respondents in the study has been 305, and it is not possible to regard these respondents as representative for the retirement club as such, and definitely not representative for retired persons in general. The results reported herein have to be regarded as tentative results, or results which have their values as compared with other results of the same kind. The empirical material has been gathered by means of a postal questionnaire, and computerized by the author.

(*) Institution of Sociology, Uppsala, Sweden.

The age range of the sample is between 50 and 93 years, with a mean of 74 years. In several earlier empirical investigations the helping patterns between retired parents and their grown-up children have been studied. In such investigations it has very early been found that the exchange is not as unbalanced as one might think. Shanas (1968) claims, referring to empirical data, that the help given is reciprocal from the children to the parents, and parents to children. The parents even help their children to a greater degree than the children help the parents. In a Danish investigation Olsen (1976) finds a similar condition. This longitudinal Danish study indicated that 72 per cent of the aged had been helping their children during the period of the investigation whereas 59 per cent of the respondents claim that they have been helped by their children during the same period. According to these investigations the pattern of exchange of help or advice is such, that the grow-up children receive more help than they give to their parents. In the Danish study, however, it has to be remembered that the respondents are fairly young (between 62 and 73 years of age) and this might be one of the reasons why the parents help their children to such a great degree.

The investigation of the members of the local Swedish retirement club differs from the result reported in the studies mentioned above. When analysing this fact, it has to be remembered that the population for study in this Swedish investigation has been rather limited, and that the number of dropouts has been great. It is also necessary to mention that a very large percentage of the members of this specific local retirement club comes from the manual labour forces. The formulations in the questionnaire have also been different in the Swedish study as compared with the above mentioned studies. In the Swedish study the respondents have been asked whether they ask their children for help or advice, and if their children address themselves to the respondents for help or advice. It has not been asked whether such exchanges of help or advice acutally have been carried out during some specific period.

Table 1. Exchange of help and advice between retired parent and grown-up children.

Exchange of help	%
No exchange at all	36
Mutual exchange	29
Only parent asked for help	26
Only children asked for help	9
	100 N = 215

Table 1 shows that among the members of the local retirement club 36 per cent of the respondents claim that no exchange at all is taking place between themselves and their children; 29 per cent claim that there is a mutual exchange between themselves and their children; 26 per cent say that only they ask for help, but the children do not ask them for help. Nine per cent claim that only the children are asking for help, while they themselves do not ask for help. Table 1 indicates that some kind of exchange of help or advice is taking place in 64 per cent of the cases. Table 2 shows that parents more often ask their children for help whereas children request help from parents less often: 38 per cent of the respondents with children claim that the children ask them for help, while 55 per cent of the parents admit that they themselves ask the children for help.

Table 2 also indicates that certain systematic differences exist between different sub-groups in this respect. The degree to which respondents say that children ask them for help is different, e.g., between men and women. A greater proportion of women report that children ask them for help; a greater number of younger respondents also

Table 2. Exchange of advice and help.

		Percentage of respondents where				N =
		children ask parents for help	parents ask children for help	mutual help exists	no exchange of help exists	
All respondents with children		38	55	29	36	236
Sex	Men	31	41	17	44	90
	Women	44	65	32	25	146
Age	- 71	56	57	41	28	81
	72 - 76	41	53	26	35	74
	77 -	17	56	19	36	76
Civil status	Unmarried, divorced, widow, widower	37	64	27	29	133
	Married	40	45	26	38	101
Economic status	Retirement allowance only	33	51	22	36	101
	Additional income	48	58	34	30	90
Health	Bad	26	66	21	28	29
	Fair	38	38	27	33	138
	Good	46	47	28	34	68

Table 2. Exchange of advice and help (cont.)

		Percentage of respondents where				N =
		children ask parents for help	parents ask children for help	mutual help exists	no exchange of help exists	
Number of friends	– 9	53	68	39	19	77
	10 –	29	42	15	46	66
Communication with neighbours	Yes	34	53	24	35	102
	No	44	54	30	33	120
Important in life	Family	47	54	33	33	107
	Amusement	38	55	28	38	32
	Care/rest	29	58	18	33	82
Hobbies	Some	53	57	37	28	100
	None	25	51	17	40	114
Activity started after retirement	Yes	60	65	45	20	40
	No	34	53	24	37	168
Travelling after retirement	More often	55	49	38	35	86
	Less often	26	59	18	33	127
Opinion on how to spend retirement	Withdrawal	24	51	17	43	93
	Activity	54	56	37	30	105

report that children ask them for help. Married respondents, those who have additional financial resources in addition to the retirement allowance only, those who regard their bodily health as good, those who have nine friends or less, those who regard family as the most important in life, those who have some kind of hobby, those who have started some kind of new activity after retirement, and those who travel more often after retirement as compared with before retirement and those who are of the opinion that the retirement should comprise of different kinds of activities also report to a higher degree that children ask them for help or advice. The general impression of this analysis is that to a greater extent children request help and advice from parents who are healthier, more active, and more outwardly directed. An exception to this rule is the fact that the children to a greater extent ask parents with fewer friends for help, in comparison with parents who have many friends.

In order to examine more closely the relative importance of these different background variables, a regression analysis has been carried out with the "children asking parents for help" as a dependent variable. This regression analysis is summarized in Table 3. It explains 39 per cent of the variance in the dependent variable, showing the most important variables to be the existence of a hobby, the age, and the number of friends which the respondent is reporting. These three variables might also reflect three different kinds of factors. The existence of hobbies might be reflecting a factor which has to do with activities. The children are then to a larger degree directing themselves towards parents who are active and alert. The age variable in its turn, can be said to reflect the factors connected with the expectations children have of aging parents, expectations which are more or less stereotyped. But it can also be that the age-variable reflects a more fundamental decrease in capacity. The fact that the number of friends is negatively correlated with the degree to which children ask parents for help, may be explained in terms of complementarity. Parents who communicate with friends to a great extent have less time for communication with children. Consequently, children ask them for help to a lesser degree. In the regression analysis shown in Table 3 one should notice that the differences in sex, as well as differences in health are of less importance when other variables have been introduced and controlled.

Table 3. Regression analysis of the degree to which children ask parents for help and advice

	Regression coefficient	Standardized regression coefficient
Hobbies	.31	.31
Age	-.16	-.25
Opinion on how to spend retirement	.11	.11
Number of friends	-.19	-.21
Communication with neighbours	.12	.12
Economic status	.12	.12
Civil status	.11	.10
Travelling after retirement as compared with before	.09	.09
Sex	.07	.07
Health	.03	.03

$R^2 = .39$

With respect to the other direction of the exchange of help and advice - the direction in which parents ask the children for help - there are not so many systematic differences found as in the case of the reverse exchange relation. Table 2, nevertheless, shows that the sex difference exists to a large degree in this case. To a much greater extent women ask their children for help as compared with the number of men who ask their children for help. Likewise the unmarried, divorced, widows/widowers, those with a bad health-condition, with fewer friends, largely ask their children for help. In this case also a regression analysis has been carried out.

Table 4. Regression analysis of the degree to which retired parents ask their children for help

	Regression coefficient	Standardized regression coefficient
Number of friends	-.25	-.26
Sex	.23	.23
Economic status	.18	.18
Travelling after retirement as compared with before	-.14	-.14
Activity started after retirement	.19	.15
Communication with neighbours	.14	.14
Civil status	.14	.14
Health	-.05	-.06
Hobbies	.05	.05
Opinion on how to spend retirement	-.05	-.05
Age	.02	.04

$R^2 = .20$

This regression analysis also indicated that the number of friends turns out to be a very important variable. Those with fewer friends ask their children for help to a much greater extent than those with many friends. Regarding the "children asking the parents for help", the sex variable was of less or no importance, but regarding the degree to which the parents request help from children, the sex variable is the second most important variable according to the regression analysis. Again it is women who mostly ask their children for help or advice. Even after other variables have been introduced and controlled, the sex variable is still very important. The regression analysis in Table 4 also shows, what is not clearly seen in Table 2, that the economic status is of rather great importance. Those with an additional income, above the retirement allowance, to a larger extent ask their children for help or advice. Although the reverse might be expected to be true, the regression analysis shows that parents with a better economic status more often ask their children for help than those parents whose economic status is low. This might be explained in terms of autonomy. Those with a fairly good economic level feel more independent, which in its turn facilitates communication with the children.

Also regarding the degree to which mutual help exists, certain systematic invariables are evident. Table 2 shows that mutual help to a larger extent is reported by female respondents, younger respondents, respondents with additional income above retirment allowance only, respondents with fewer friends, respondents who have certain kinds of hobbies, those who have started some specific kind of activity after retirement, those who travel more often after retirement and those who are of the opinion that retirement should be devoted to different kinds of activities. The profile of the respondents where mutual help exists is very much like the profile where the children to a large extent ask parents for help. It is the respondents who are alert and directed outwards who to a larger extent report mutual exchange of help and advice between themselves and their children. In order to find out the relative importance of different factors, also the degree to which mutual help exists has been the subject of regression analysis. In this connection Table 5 shows the most important factor to be the existence of some kind of a hobby. The factor next in importance is the number of friends reported by the respondent. Also in this case the number of friends reported is negatively correlated to the degree to which mutual help between parents and children exists. The mutual help

Table 5. Regression analysis of the degree to which mutual exchange of help and advice exists

	Regression coefficient	Standardized regression coefficient
Hobbies	.23	.24
Number of friends	-.29	-.23
Sex	.18	.19
Economic status	.18	.19
Age	-.10	-.17
Civil status	.15	.15
Communication with neighbours	.11	.12
Activity started after retirement	.09	.08
Opinion on how to spend retirement	.06	.07
Health	-.05	-.07
Travelling after retirement as compared with before	.04	.04

$R^2 = .34$

exists where the respondents have fewer friends. The variables of sex and economic status comes third, after which follow the age variables and the variables of civil status and communication with neighbours. Concerning the existence of mutual help and advice, it should be noted that the regression analysis shows that factors like the condition of health and the opinion on how to spend the retirement are of no importance after other variables have been introduced. Table 2 also shows that certain systematic invariances are found concerning those cases where no kind of exchange of help or advice exists. The percentage of respondents reporting no exchange at all is highest amongst the men, among those who report having a large number of friends, among those who have not started any kind of new activity after retirement, and among those who are of the opinion that retirement should be a time of withdrawal and rest.

Table 6. Regression analysis of the degree to which no exchange of help or advice exists

	Regression coefficient	Standardized regression coefficient
Number of friends	.24	.27
Communication with neighbours	-.15	-.15
Hobbies	-.14	-.14
Economic status	-.13	-.13
Sex	-.12	-.12
Civil status	-.10	-.11
Activity started after retirement	-.10	-.08
Travelling after retirement as compared with before	.09	.10
Age	.04	.06
Health	-.03	-.03

$R^2 = .17$

In Table 6, the regression analysis of the degree to which no exchange of help or advice exists, it is shown that the reported number of friends is the most important dependant variable.

Table 2 reveals some tendencies of a more general kind. One such general tendency is that the exchange of help tends to be greater among the female respondents. This primarily results from the fact that the female respondents themselves address themselves to the children to a larger extent than the male respondents do. It is not a result of the fact that children ask mothers for help to a significantly greater extent than they ask their fathers. In other words, when children ask for help or advice, they are not acting according to traditional sex roles, but when parents ask children for help, normally they are acting in terms of traditional sex roles.

Another interesting finding is that the age variable is of quite different importance for the different directions of the exchange relationship. When we study the degree to which children ask parents for help, the tendency decreases with the increasing age of the parents, but when we study the reverse relationship, the degree to which the parents ask the children for help, this tendency is independent of the age variable of the respondent. As a result of the fact that the childrens' tendency to ask parents for help is decreasing with the increasing age of the parents, there is also an age dependent relationship concerning the degree to which mutual help and advice exist. With the increasing age of the respondent, the degree of mutual help decreases. The economic status of the parents has a similar effect to that of the age variable. Children to a larger extent have recourse to parents with a good economic standing and the degree of mutual help is also broader in this case.

From Table 2 we can deduce that the health condition is related to the exchange of help and advice in a way that might be expected. With respect to the degree to which children ask parents for help, the tendency is greater the healthier the parent is. Concerning the exchange relationship in which the parents ask children for help, the tendency is greater the worse the parent's health condition is. The regression analysis of these two findings, however, shows that the health variable, as such, does not have as great an importance as one might believe from the study of Table 2. Other variables, which are correlated to the health variable, are of more importance. This appears from the fact that the health variable receives very small values on the regression coefficients of the regression analysis.

Another general finding, which pervades the whole study is the fact that the number of friends reported by the respondents is significantly related to the degree of exchange between parents and children. The degree of exchange is always larger where the parents report having fewer friends as opposed to those who report having a larger number of friends. As already mentioned, this might be explained in terms of complementarity. The exchange of help and advice either occurs between parents and children or between parents and their other friends.

Another very general finding is that the children to a larger extent address themselves to parents who are active, alert and directed outwards. Such differences, however, do not affect the degree to which parents adress themselves to the children for help and advice.

In cases where exchange of help or advice exists between parents and children, the respondents have also had the possibility to specify which kind of help or advice they generally receive. Although relatively few of the respondents have availed themselves of this possibility, those which have been registered give an idea of the specific kind of help or advice that is predominant. Table 7 shows that children often appeal to parents for help with child nursing, when seeking general advices and occasionally when requesting economic support.

Table 7. Kinds of help or advice, asked for by the children

Kinds of help or advice	Number
Child nursing	17
General advice	18
Economic support	14
Advice concerning children	3
Cleaning and housework	5
Help in connection with illness	1
Unspecified help	26

Table 8, on the other hand, indicates that parents tend to request help from their children primarily in time of illness. The most common requests are those which entail help with housework and cleaning.

Table 8. Kinds of help or advice, asked for by the retired parents

Kinds of help or advice	Number
Help in connection with illness	32
Cleaning and housework	12
Shopping	6
Help to fill in forms etc.	6
Economic support	3
Gardening, repairing	4
General advice	14
Unspecified help	47

REFERENCES

Olsen, H., et al. *Familiekontakter i den tidige alderdom*, (Family Contacts in the Early Stages of Old Age), Socialforskningsinstituttet, København, 1976.

Shanas, E., Family Help Patterns and Social Class in Three Countries, in Neugarten B.L., ed., *Middle Age and Aging. A Reader in Social Psychology*, University of Chicago Press, Chicago, 1968.

15. The geriatric ward and the family - a study of communication

Lena Dahl (*)

INTRODUCTION

The problem of information in medical care has lately come into focus. It is desirable to achieve a thorough knowledge of health care, illness, and medical care by means of increased information. A lack of knowledge may result in overconsumption of both institutional and non-institutional care. Information is a problem not only in the relation between personnel and patient but also in the relation between personnel and relative as well as between relative and patient. Exchange of information is of great importance for an effective process of rehabilitation.

The information about illness and medical care is particularly important in the long-term rehabilitation process. An effective rehabilitation requires great resources of personnel as well as support from the patient's family. The aged patient must not be seen as an isolated person. Though temporarily he belongs to the hospital, he is still bound to his home and to his family. "Family" is here defined as a bigger unit than the nuclear family and includes, for instance, grandchildren and other close relatives.

(*) Institutet för Gerontologi, Jönköping (Sweden).

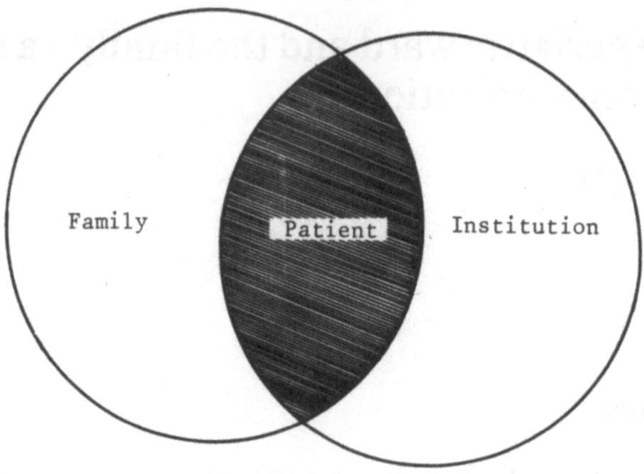

Figure 1. The patient in institutional care remains a family member.

In Sweden social reforms have made it possible for old people to live economically independent of children and other relatives. The need for affective relations and mutual responsibility, however, has not decreased as a consequence of these reforms.

The hospitalization of an old person often involves both physical and psychological stress. It is important in this situation that the patient receives emotional support from his family. The family becomes an information mediator of great significance and gives the patient an opportunity to keep in touch with the world outside the hospital. When the patient leaves the hospital for rehabilitation care outside the hospital, the role of the family becomes even more important. There ought to be a continuity between the proceedings taken by the hospital and the proceedings that will be taken when the patient is back home again. The family is a natural link between the hospital and home and takes the responsibility for the realization of the various rehabilitation proceedings.

Informing the patient about his/her illness, medical treatment and prognosis is one step of the rehabilitation process. With adequate knowledge the patient himself/herself can take an active part in the recovery. To the family, illness and hospitalization means a state of crisis including changes in daily life. Responsibility for and worries about a near relative together with a lack of knowledge about the illness and its course affect the possibilities of the family with regard to the help and support. In this situation it is important that the family receive good advice and help from the nursing personnel and that the family supply information about the patient to the personnel. What is the purpose of family information? It intends primarily to improve the knowledge of the illness, old age changes, hospital care and routines, as well as to affect the attitudes of the family towards the rehabilitation process and the nursing personnel in a way that will enable the family to take an active part in the process itself. Due to professional secrecy the informative activity creates a dilemma to the personnel. It is mainly the doctor who is responsible for the information. He shall, in agreement with science and reliable experience, supply the patient with the advice and if possible with the treatment required by the condition of the patient (SFS 1963:341). Moreover, the patient has the right to know what is written in the case-record. Inquiries about a patient should be answered by the nurse in charge of the ward where the patient is nursed. The doctor can take it for granted that he has the patient's permission to inform other doctors about the case. This is also valid with respect to informing the patient's family. The patient should, however, be entitled to forbid supplying information to other doctors as well as to members of the family. The doctor has to decide from case to case what information should be given to the patient and his family.

In Sweden, a recommendation for geriatric care stresses the fact that it is important to establish good contact between the close relatives of the patient and the personnel in the long-term care. Nevertheless, systematic information of this kind does not exist to any great extent in Sweden. It was precisely in order to evaluate the existing exchange of information between members of the family of the patients in geriatric wards and the personnel that this explorative study was undertaken. This study is the first step to a more extensive investigation about the mutual exchange of information between the family and different categories of personnel on the nursing team. Diverse problems have been studied, e.g.; information to the family, information to the personnel, need for information to the personnel, source of information, method of information and time for information.

MATERIAL AND METHODS

All members of the day-time personnel and relatives of patients of four different wards of a long-term care (geriatric) hospital constituted the material of this study. The personnel group totalling 58 persons was represented by all members on duty at the time of the study. Of these, 55 persons altogether participated, which makes a participation rate of 94.8 per cent. The categories represented were : registered nurses, nurses, nurse's aids, occupational therapists and physiotherapists. The majority of the personnel group consisted of registered nurses, nurses and nurse's aids. The average age was 30.3 $\pm$ 11,2 years (M $\pm$ S.D.). 58 per cent were under 30 years of age and only 44 per cent had been working in medical service for more than 4 years while 36 per cent had been working in geriatric care for more than 4 years. The relatives, consisting of 71 persons were those whom the patients at the time of admission into the hospital had listed. Of these 50 persons participated in the study, which makes a participation rate of 70.4 per cent. Altogether 21 persons did not take part: 14 relatives did not respond and 7 had moved to addresses unknown to the hospital. Of the 50 participants 25 per cent were in a wife/husband relationship to the patient, 27 per cent were children of the patients and 33 per cent were related to the patients in some other way. The mean age of the patients was 79.9 years and 58 per cent belonged to the 80-94 year age group. 48 per cent of the patients had former experience as patients while 86 per cent of the relatives had no professional experience of hospital care.

For the two groups two slightly different questionnaires were used, one for the personnel and one for the relatives. The questions were of a multiple-choice type, questions of yes/no type as well as open questions requiring more detailed information. The questionnaires were answered by the personnel during work-time and the relatives received and returned their questionnaires by post.

RESULTS

Before the patient was hospitalized, 52 per cent of the family members already had some knowledge about the patient's illness. At the admission of and in some cases afterwards, 82 per cent stated that they had been informed of the nature of the illness by doctors or other categories of personnel. In addition, 31 per cent got their knowledge from TV, radio or newspapers and 24 per cent from friends or books. Among the relatives 32 per cent thought that they had received good information about the illness and 26 per cent felt they had satisfactory information about the medical treatment.

The information booklet distributed by the hospital, which gives advice and information about hospital routines, is an existing form of patient/family information. However, 65 per cent of the relatives did not know of this booklet, and only 31 per cent had read it.

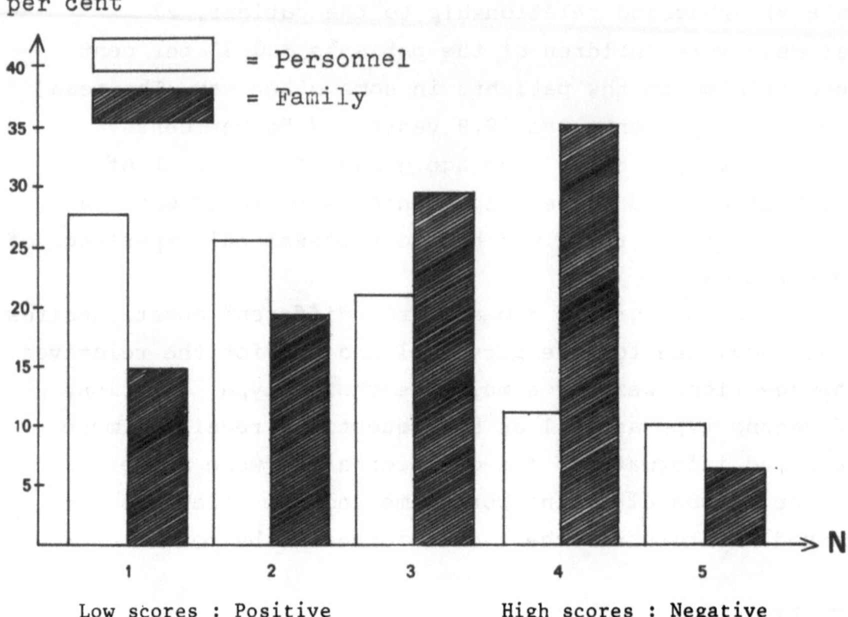

Figure 2. The exchange of information about the patient's disease: a comparison between the personnel and the family.

The major part of the personnel and the relatives agreed that relatives need information about the illness and medical treatment of the patient. Almost as many people in the same groups thought that relatives need information about the patient's daily life in the hospital. Of the personnel 53 per cent thought that the relatives generally were informed about the illness while 45 per cent agreed that the same was true with respect to medical treatment.

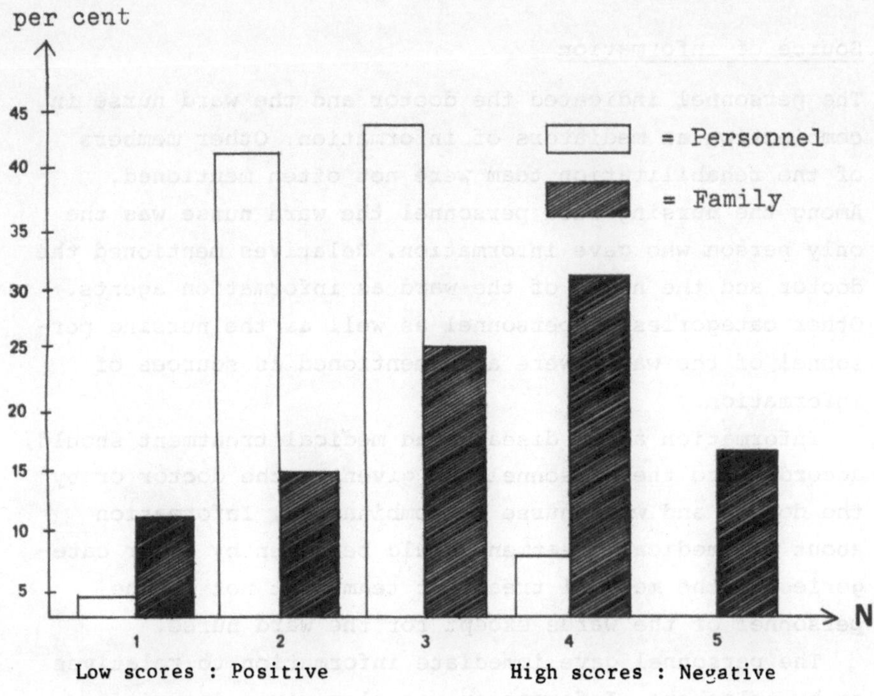

Figure 3. The exchange of information about the patient's treatment: a comparison between the personnel and the family.

Half of the personnel group was doubtful about the sufficiency of the information given to relatives with respect to the disease and medical treatment of the patient. Almost all members of the personnel group and the group of relatives agreed that the personnel needs to be informed about the patient by the relatives. Only 31 per cent of the personnel were of the opinion that they had been informed about the patient by his family, while 38 per cent of the relatives stated that they inform the personnel about the patient. The information that the personnel received from the family was, however, regarded as sufficient by 36 per cent of the personnel.

Source of information

The personnel indicated the doctor and the ward nurse in combination as mediators of information. Other members of the rehabilitation team were not often mentioned. Among the nursing ward personnel the ward nurse was the only person who gave information. Relatives mentioned the doctor and the nurse of the ward as information agents. Other categories of personnel as well as the nursing personnel of the wards were also mentioned as sources of information.

Information about disease and medical treatment should, according to the personnel, be given by the doctor or by the doctor and ward nurse in combination. Information about the medical treatment could be given by other categories of the medical treatment team, but not by the personnel of the wards except for the ward nurse.

The personnel gave immediate information to relatives or by telephone. Information was also given in writing. Relatives said that they were informed personally by the personnel, or from time to time by telephone. The personnel wished to inform the relatives face to face. While a combination of existing ways of information should be used, relatives prefer to be informed personally.

Time for information

Most relatives have been informed by the personnel when the patients came to the hospital ward and some when the diagnosis was completed or when changes in the patient's condition occurred. Information about the medical treatment has mostly been given continually or at the time that the appropriate treatment was to be administered.

Relatives stated that they had received information during the period that the patient has spent in the ward and sometimes before the patient was hospitalized. The majority of the personnel thought that information should be given when the patient comes to the ward, or when the diagnosis is evident or when the patient is worse. Information about the medical treatment should be given at the time of hospitalization or at the time of treatment.

Satisfaction with the exchange of information

Fifty per cent of the relatives were satisfied with the information they received from the hospital personnel. One third of the personnel were satisfied with the existing exchange of information.

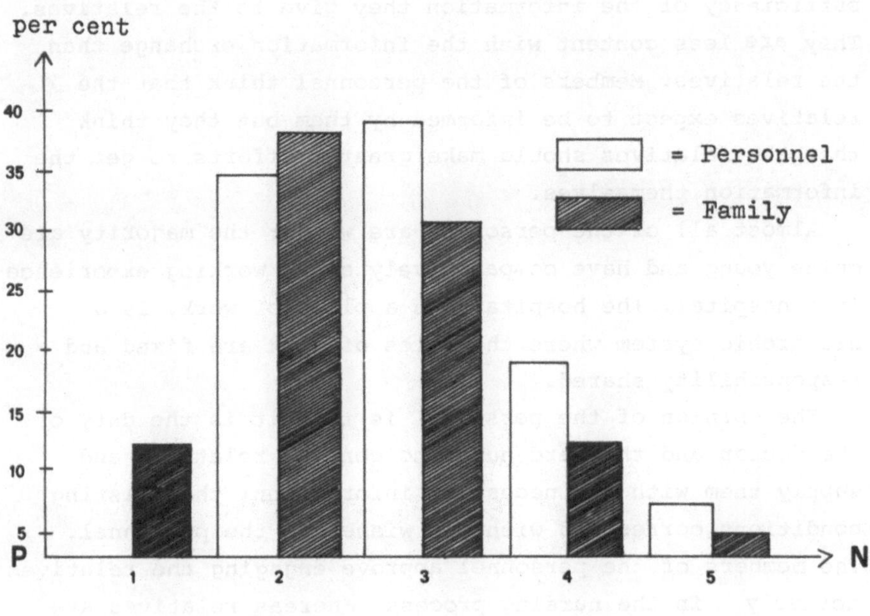

Figure 4. The satisfaction with the information exchange: a comparison between personnel and family.

The majority, that is, 3/4 of the relatives stated that they themselves had to ask for the information they wanted. Half of the personnel were of the opinion that the relatives do not have to ask for information about the patients.

Nearly all of the personnel thought that the relatives expect information about the patient from the personnel. Just as many thought that it should be to the interest of the relatives to obtain the information they want.

DISCUSSION

The personnel members think that the relatives need information and are of the opinion that the personnel should inform the relatives spontaneously without the relatives asking for it. The personnel are doubtful about the sufficiency of the information they give to the relatives. They are less content with the information exchange than the relatives. Members of the personnel think that the relatives expect to be informed by them but they think that the relatives should make greater efforts to get the information themselves.

Almost all of the personnel are women; the majority are quite young and have comparatively short working experience in a hospital. The hospital, as a place of work, is a hierarchic system where the roles of work are fixed and responsibility shared.

The opinion of the personnel is that it is the duty of the doctor and the ward nurse to contact relatives and supply them with the necessary information; the existing conditions correspond with the wishes of the personnel. The members of the personnel approve engaging the relatives actively in the nursing process. Whereas relatives are occasional visitors at the hospital, contact with the personnel is not evident. Relatives do not expect to get information nor do they feel that they actually do get information to the extent claimed by the personnel members. Relatives feel there should be a reciprocity in the exchange of information, viz., that personnel provide relatives with information about the patient's condition and that they in turn provide the personnel with the necessary information about the patient. Usually it is relatives who contact the personnel hardly realizing that the personnel feel the need of approaching the family.

Relatives are comparatively satisfied with the existing exchange of information. This information is not difficult to understand, and the relatives consider themselves positively treated by the personnel. Generally, the relatives are satisfied with the existing conditions. Wishes regarding the method and the time for dispensing information correspond with the actual conditions. The relatives accept the existing system of hospital care.

The informative brochure could be a good kind of systematic information but does not serve hoped for the purpose. The majority of the relatives do not know of its existence. On the whole, the relatives are content with the existing system of information regarding the long-term wards studied. Nevertheless the personnel are not so satisfied as the relatives; they feel the need for improving the possibilities of dispensing information about the patients entrusted to their care.

ABSTRACT

In order to evaluate the exchange of information between the relatives of the patients and the personnel a questionnaire study was conducted in a long-term care (geriatric) hospital. Of the 58 members of the personnel, 55 participated (94.8 per cent). Among the relatives 50 out of 71 participated (70.4 per cent).

There was a slight difference between information received from relatives and information given by the personnel. The personnel were of the opinion that they had given information to a greater extent than the relatives admitted receiving. Both groups were convinced of the need for exchange of information. The source of information was the doctor or the nurse in charge of the ward. Whereas relatives were quite satisfied with the existing exchange of information, they admitted that they were obliged to ask for it themselves. The personnel were of the opinion that relatives should make an effort to obtain by themselves the information they demanded. They did not consider the opportunities of exchange satisfactory.

Research and methods

16. Encounter groups with elderly persons: a supplement to the familial support systems

Nan Stevens and Michel Wimmers (*)

Since the fall of 1974, staff members and graduate students in the department of social gerontology at the University of Nijmegen in the Netherlands have been conducting encounter groups with elderly persons. The wish to test several assumptions motivated us to begin experimenting with these groups. Our assumptions are that:

1. personal growth is possible late in life;
2. personal growth involves change in the direction of acquiring more insight in oneself and in the behaviour of others and putting this insight to use;
3. personal growth can be stimulated by an encounter group experience;
4. elderly encounter group members are able to describe their own experience in the group, including any change which might have taken place in their own behaviour and/or experience as a result of participation (v.d. Boom and Stevens, 1976).

The elderly persons who participate in the groups do so voluntarily; originally participants were recruited in service centers for the elderly via the social workers who worked there. Our original target group was healthy older people, living independently in the community. A subsidy for taxi fares has enabled us to include less mobile elderly, including those residing in homes for the elderly. Our

(*) Department of Social Gerontology, Nijmegen (The Netherlands).

recruitment channels have broadened to include referrals by ex-members, family doctors, and directors of housing for the elderly, as well as articles in neighbourhood newspapers. Up until now, approximately eighty older people have participated in the groups; twenty have participated in a group for the second time. The group members' motivation for participating often include a desire for new learning experiences and for new social contacts, the need to find new purposes in life or new roles for oneself (e.g. after the loss of one's partner or after retirement), or the need to survey and reintegrate previous experience, often troublesome, which has resurfaced with old age.

Ideally a group consists of six to eight members; usually there are two leaders and one or more observers. The session are tape recorded so that the team of leaders and observers responsible for a group can monitor its progress and their own functioning. The number of sessions is jointly determined by group members and leaders; usually a group meets for 15-20 sessions on a weekly basis. Most of the groups have met at the psychology department of the university; however two groups have preferred to meet in a service center in the neighbourhood in which the members live. The leaders do not work with an agenda or predetermined topics for discussion but prefer to let the members decide what they want to talk about with emphasis on relating their own personal experiences. In this way the most salient themes and experiences of group members in their present situation are discussed. Themes which are frequently brought in the groups for discussion include a member's relations with adult children, with his/her partner, sexuality, living alone, living in a home for the elderly, physical handicaps and health in general, death in the family or among aquaintances, one's role in neighbourhood activities.

The role of the leaders is primarily to structure conversation in the group so that members listen to one another, to limit interruptions, to encourage members to speak personally instead of generally, to identify the linkage between the statements of different members and convey this to the group. In addition to this structuring, each leader develops his/her own style of leadership, which varies in the amount of confronting, supporting, understanding, and advising in relation to group members. Some leaders choose to focus on individual group members while others are more group-centered. They also differ in the extent to which they bring their own personal experience in the group.

The students who co-lead the groups along with staff members are often majoring in gerontology; all of them have a special interest in working with the elderly. Before beginning with the groups of elderly they have received several months of training in group work. It is perhaps important to add that most of the graduate students are in their twenties. This means that there is a considerable age and generation gap between the leaders of the group and the members who are in their sixties, seventies and eighties. Caplan advocates the development of such programs in which the older and younger generations "can learn to respect other, can learn each other's language, and can discover and appreciate the other's capacities and potential helpfulness" (Caplan and Killilea, 1976, 33).
It is a remarkable experience for younger persons working in such a group to realize that the age difference seems to disappear once the group reaches the stage in which members and leaders freely express themselves and are actively involved with helping one another.

It is this last point, that of the mutual help between group members which we wish to emphasize in this paper. "Older persons can give help and they can receive help" reads one of the specifications of the theme of this congress. In every encounter group, there are several members who show a natural and spontaneous capacity for dealing with pain and suffering of others (Rogers, 1973, 28). Rogers describes this phenomenon as the "development of a healing capacity in the group". A recently widowed woman feels especially understood by a woman, who has recovered from her bereavement at the loss of her husband, when she describes how hard it is for her to learn to be alone. A mother who does not understand her daughter's decision not to have any children is helped by other members of the groups, especially by those who have had to struggle to accept the changing attitudes and life styles of their children and grandchildren. Another man, going through the painful process of divorce at 63, is especially sensitive to the pain and suffering of other group members in various situations. In one group, members encourage a man to accept an office in a senior citizens organization when it is clear that his hesitation is due to a lack of self-confidence, to the feeling that he is "too old". At the end of the group, when he has been elected as chairman, he thanks the group members for helping him get his selfconfidence back. These are a few examples of how a healing capacity develops and works in an encounter group, in this case, with elderly persons.

Another phase in encounter groups which Rogers describes in the development of helping relationships outside the group sessions. This is perhaps most apparent in groups in which the members come from one neighbourhood. Members who admit to being lonely begin to visit one another, often making arrangements to meet while at the weekly group session. When a woman complains that her doctor only sees her long enough to write a prescription for her, another woman offers to introduce her to her own doctor who takes the time to listen to his patients. If a person is troubled by something between sessions, he or she may call up another group member to talk, knowing that he/she will be a sympathetic listener.

In so far as elderly people develop mutual helping relations within these groups, the groups may be serving a broader function in the community; a successful group can supplement a group member's family by acting as an additional support system. Caplan defines a support system as a "continuing social aggregate which provides individuals with opportunities for feedback about themselves and for validation of their expectations about others". (Caplan and Killilea, 1976, 19). There is evidence that the presence of support systems is related to a lower incidence of physical and mental disease, hence support systems can be seen as a kind of buffer against disease. Caplan suggests that this is due to the fact that in these relationships the person is dealt with as a unique individual; others are interested in him in a personalized way, they are sensitive to his needs which they deem worthy of respect and satisfaction. This kind of personal involvement also characterizes our groups of elderly. Furthermore several functions which a group of elderly can serve correspond to those of a support system; a group can (1) provide information about the world in the form of personal experiences reported there; (2) members can help one another make valid assessments in bewildering situations; (3) the group can serve as a guide and mediator in problem solving by encouraging members to talk freely about personal difficulties, by offering advice based on similar experience and by providing information on external care and assistance, and (4) the group can serve as validator of identity by affirming members self worth, recognizing members strengths and skills, and through the expression of solidarity. While participation in a group is not a substitute for the support system provided by the family of an elderly person, the group often does become an auxiliary support system for many members. Several elderly persons who participated in different groups report that while they were in the groups they discovered that they were dissatisfied with aspects of their own behaviour with family members; with the helps of insights gained in the group, they report that they are working at changing their behaviour and that specific relations with family members have improved. Breaking out of old patterns of behaviour and trying out new behaviour is perhaps easier with a group of supportive peers than it is in one's family where roles and rituals are more confining. An elderly man who expresses romantic interest in a female group member may be hesitant to discuss his longing for companionship with his children. A woman feels free to talk about her anxiety when she is alone at night in the group, while she does not want to worry her children by "complaining". With the support of the group as well as her family, she begins to take steps to try to master her anxiety, by seeking professional help and by changing her living situation.

In closing, we wish to make several points:

1. Our original assumptions have been supported by groups members' own accounts of their experience in our groups, i.e. that personal growth is possible late in life. However, our attempts to measure the effects of participation in the groups have had mixed results; we are still in the experimental stage regarding research on the groups.
2. Nevertheless the reactions of group participants have encouraged us to continue with the groups; after four years, we can no longer claim to be in an experimental phase. The groups have proved their worth. At the same time we feel that the organization of the groups should be taken over by a community agency in order to reach more elderly persons. The most attractive setting for the groups appears to be the neighbourhood service center for the elderly; an alternative is the mental health agency which unfortunately in the past has not served many elderly clients.
3. The older person's relative unfamiliarity with group work and other forms of psycho-therapy as well as a

general hesitancy to seek psychological help, necessitate an active recruitment program if one wishes to organize groups of elderly. Ex-members of groups are often the most effective recruiters, in that they are able to describe possible benefits of participation based on their own experience.

4. The following step might be to train ex-members to co-lead groups of elderly as is done at the Continuum Center for Adult Counseling and Leadership Training at Oakland University in Michigan (Waters, Fink and White, 1976). One woman who took part in our groups has recently started a group in her neighbourhood with a social worker.

ABSTRACT

A program run by the department of social gerontology at the University of Nijmegen in Holland offers elderly persons living independently or in homes, the opportunity to participate in an encounter group, such as those developed by Carl Rogers.

The goals of the group, working methods, the role of graduate students, group members' motivations for participating and themes discussed are described in this paper. Emphasis is placed on the mutual helping behaviour by group members, both within and outside the group. These groups can serve as support systems which supplement an elderly person's familial support system, by providing information about the world, helping members make assessments of difficult situations, affirming members' self-worth, expressing solidarity and helping with the solution of problems. In this way the groups serve a preventive function in mental health care of the elderly and should become part of community services for the aged.

REFERENCES

Caplan, G., and M. Killilea, *Support Systems and Mutual Help: Multidisciplinary Explorations*, Grune + Stratton, New York, 1976.

Rogers, C., *Encounter Groups*, Penguin Book Lts, Middlesex, England, 1969.

Van den Boom, C., and N. Stevens, Three Elderly Encounter Group Members, doctoraal thesis, University of Nijmegen, May 1976.

Waters, E., S. Fink, B. White, Peer Group Counseling for Older People, in *Educational Gerontology*, 1, 1976, 157-170.

17. The relation between elderly parents and their children: a two-sided research programme of the awareness-context of this intergenerational relation

Anton Bevers (*)

INTRODUCTION

On consulting studies on the social relations of elderly people, one can conclude that the children play a very important part in the social network of the married elderly people. One can also establish that in most studies the structural aspects or rather the sociological form-aspects of this relation are accentuated: purpose, length and frequency of visits, forms of help, living accomodation and living distance, family members, social surroundings etc. In the third place one might observe that the information on the social relations of the elderly mostly comes from the elderly themselves, that is to say, from conversations with the elderly about their social relations and not e.g. with their children. This research programme also places the relationship between elderly people and their children in a central position. In this programme however, we are mainly interested in the material aspects of this relation, although much attention will be given to the sociological form-characteristics.

The intention of this research is two-sided; we want to investigate the subjective engagement of both parents and children. We can speak of a meaningful social relation when partners know how to adjust their attitude among themselves with regard to the other, that is to say, when

(*) Gerontological Center, University of Nijmegen.

they can put themselves into the other's position. Therefore an important question in our research is: to what extent and on what points of the relations have the elderly parents and their children mastered the techniques of 'role-taking' ? Proposing questions in this way requires a two-sided research. The possibility of putting oneself into the position of the other is promoted by what we know about the other. Every social relation is therefore also strongly determined by the knowledge partners have of each other. In our research therefore, we shall have to propose the question as to what parents and children do and do not know about each other: we are then asking after the awareness-context of this relation. The term 'awareness-context' was first used by Glaser and Strauss. In their research on interaction-patterns and -processes round dying patients in a number of hospitals, they give the following definition of awareness-context: "the total combination of what each interactant in a situation knows about the identity of the other and his own identity in the eyes of the other". By participant observation method they arrived at four main types of awareness-context: closed awareness-context; suspicion awareness-context; mutual pretense context and open awareness-context. The question as to what actors in a social relation do and do not know about each other, belongs to the sociology of ignorance. In this connection the works of the German sociologist and philosopher Georg Simmel (1858-1918) must certainly be mentioned. Ignorance is according to Simmel a fundamental characteristic of social relationship: every social relation is characterized by what individuals do and do not know about each other. In his sociology of the secret Simmel has given detailed information on the practical function of ignorance in everyday life. Ignorance, just like knowledge and truth, is available to our practical necessities of existence. Simmel attaches pragmatic values to ignorance. In the contact between people it applies "dass wir nicht nur so viel Wahrheit, sondern auch so viel Nichtwissen bewahren und so viel Irrtum erwerben, wie es für unser praktisches Tun zweckmässig ist" (Simmel, 1968, 258). Simmel deals with a great number of forms and form-characteristics of ignorance in his sociology of the secret: he discusses the role of the lie and of trust, forms of secrecy and attempts at exposing, the typical character of intimacy and discretion, differences between a stranger, an acquaintance and a friend.

The relation between knowing and not knowing about each other can be controlled in several ways. In the symbolic interactionism these interactional processes have been closely observed and analysed. What we know and do not know of each other, but also of ourselves, is brought about via communication processes and specially via the role-changing mechanism, called 'role-taking' by G.H. Mead. Cooley's concept 'looking-glass self' is related to the same social interactional mechanism. In the symbolic interactionist perspective, three elements of social relations in their mutual correlation are inquired: one's own identity, the other's identity and one's own identity in the eyes of the other. Social relations can be distinguished from each other by the degree in which the people concerned have knowledge of each other's identity and of one's own identity in the eyes of the other. (cfr. types of awareness-context of a social relation according to Glaser and Strauss). Consciously or unconsciously, people always act without complete certainty that they know each other's true identity. Forms of concealing, revealing, pretending, suspecting, openly declaring or twisting of the facts are present in everyday relations, e.g. between buyer and seller, husband and wife, writer and reader, teacher and pupil, parents and children, etc.

Next to the studies by Simmel, Mead, Cooley, Goffman, Glaser and Strauss, a.o., referring to social relations and mutual engagement of partners on each other's intentions, numerous other researches have been made on the

presence or absence of agreement between the partners with reference to the aspects of this subjective engagement. These concensus-studies mainly look for an agreement of opinion between people, specially with reference to those relations in which presence of consensus is considered to be functional. If also the question is asked to what degree the partners are aware of each other's opinion, then we can say that the concensus-study has developed to include the awareness-context of the investigated relation. From two-sided, and also from one sided research on the relation between elderly parents and their children appears the asymmetrical character of these relations with regard to the degree of mutual dependence, affective engagement, role and position in the family, composition of social network, activities, living accomodation, financial position and health. The central question of our research is related to what the parents and their children do and do not know about each other: to what degree does this relation of knowing and not-knowing belong to the characteristics that define the assymetrical character of this relation?
One of the conditions for the research for asymmetry in social relations as to what the people concerned know about each other - the awareness-context - is a two-sided research. (with two-sided we mean here: both parents and children participate in the research). To obtain more insight in the qualitative aspects of this intergenerational relation, the subjective world of experience of the persons concerned is an indispensable source of information. Through such a two-sided research, the research-worker finds himself in the exceptional position of getting knowledge of characteristics of this relation which remains concealed for the people concerned.

THE RESEARCH PROBLEM

The general question contains the following five parts:

I. 1. What subjective meaning do elderly people and their children attach to aspects of their mutual relationship?

2. Is there a difference between them in this respect?

3. Does asymmetry occur, that is to say, if there is a difference does it clearly point in a certain direction?

II.4. To what degree do elderly people and their children know each other's subjective engagement?

5. Is there a difference between them in this respect?

6. Does asymmetry occur, that is to say, if there is a difference does it clearly point in a certain direction?

The first three questions refer to the difference or agreement of opinion between parents and children. With the help of the answers to the first three questions, the next three questions give insight into the types of knowledge of each other's opinion. Next to the question on the difference/agreement of opinion between parents and children and the mutual knowledge thereof, there is a third element from which the asymmetrical character of this relation may appear. The question to be put with reference to this matter is:

III. 7. How do elderly parents and their children deal with the knowledge they have about each other?

What forms of communication and behaviour are chosen or avoided by them based on what they do and do not know about each other and are these forms meant or not meant to bring changes in an existing relationship of (not)knowing? The question is mainly to find the right interpretation of the attitude of parents and children towards each other; after this one can decide whether one can speak of open communication, closed communication, communication based on suspicions or pretence. It is possible and even self-evident that when the contents of the relations vary, the form of communication will vary too.

Up till now we have formulated three problem-areas: 1. difference/agreement of opinion. 2. knowledge of each other's opinion and 3. forms of communication in order to maintain or manipulate information about each other. These three parts together belong to the awareness-context of the relationship between elderly parents and their children.

The information gathered with reference to these three questions must be investigated as to their correlation to the following variables:

IV. 8. What is the correlation between the personal characteristics of parents and children and the three elements of the awareness-context (difference/agreement of opinion - knowledge - communication pattern) ?

To the personal characteristics are considered to belong: marital status, age, occupation, education, old age experience, images of old age.

9. What is the correlation between relation-characteristics of parents and children and the awareness-context?

To the relation characteristics are considered to belong: form and contents of mutual visiting, assistance and living-distance.

10. What is the correlation between relation-network-characteristics of parents and children and the awareness-context?

To the relation-network characteristics are considered to belong: members of the family, form and contents and composition of other social relations.

It will be obvious that the above questions (8, 9, 10) are exclusively meant to place the data of the awareness-context in a wider sociological frame. The awareness-context of the 'face-to-face' relation has a social-structural context to which, amongst others, we must reckon the individual social network of the persons concerned. For example we want to know from the elderly parents and their children whether they are family-minded in their social relations or if they are not: in other words, we ask if there is asymmetry in the composition of the social network of parents and children.

V. 11. Finally we ask ourselves to what extent the data of the awareness-context, and especially the degree of asymmetry with reference to difference of opinion and knowledge thereof, can be interpreted in the context of the theories on nuclear family and the position of the elderly in modern society.

Schedule of problems

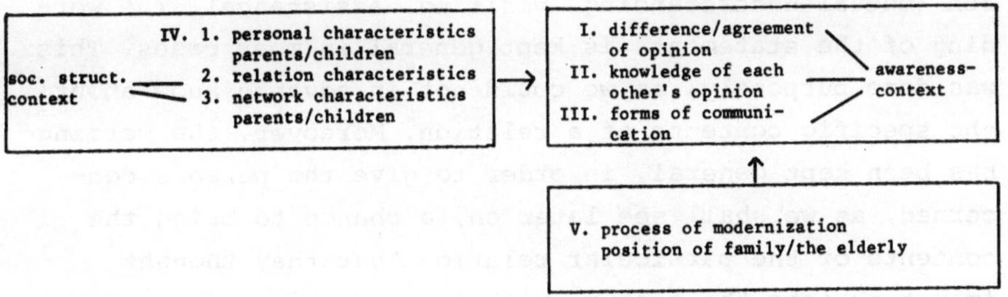

Operationalizing of the research-problem

Based on literature study we have formulated 19 statements for the research, each of which, we presume, refers to an aspect of the parent/child relation and of which the answers can provide us with more understanding of the subjective engagement of parents and children in their mutual relation. The statements refer to mutual visiting, assistance, living situation of parents, mutual understanding, education of children, ideas on religion/politics (for list of statements see appendix).

The contents of the items vary strongly. Sometimes the statement only refers to the situation of parents (living situation, leisure time, financial position), sometimes it only refers to the situation of the children (education), in other cases the statement refers both to parents and children (mutual understanding, visiting, assistance). The wording of the statements is kept general in most cases. This was done purposely, as we could not in advance know about the specific contents of a relation. Moreover, the wording has been kept general, in order to give the persons concerned, as we shall see later on, a chance to bring the contents of the particular relation that they thought important into the picture with some examples.

The statements do not form a scale with which a total score, as the extent of consensus between parents and children, can be made. The answers to the items cannot be added up via a score-awarding. The items are too dissimilar for this, as regards the contents and also the difficulty in connection with 'role-taking' and because of the variation of the number of possibilities of answers per item; this means that one must consider per item how parents and children have answered. The fact is that we are not interested in establishing how large the difference of opinion is between parents and children on the whole relation: adding up is meaningless, nor it is our intention to do so, in view of the contents. It might however be possible to deal with items of similar tendency as one material theme.

The specific questions with reference to the awareness-context are therefore:

- on what material aspects of the relation does difference/agreement of opinion exist?
- between which parent/child dyad?
- on which contents of the relation do they or do they not know each other's opinion?
- who is most capable of taking the role of the other?
- is difference of opinion known more often than agreement?
- who see more differences or agreements?

In order to establish difference/agreement of opinion and knowledge thereof, we need the following information:

1. opinion of parents, according to parents themselves----- symbol P
2. opinion of children, according to children themselves--- symbol C
3. opinion of parents, according to children-------------- symbol PC
4. opinion of children, according to parents--------------- symbol CP

The list of questions for the parents gives us information P and CP. The list of questions for the children gives us information C and PC. In order to study the awareness-context, the following comparisons are of importance:

1. P - C objective difference/agreement of opinion
2. P - CP subjective difference/agreement of opinion acc. to parents
3. C - PC subjective difference/agreement of opinion acc. to children
4. P - PC children know their parents' opinion: yes/incorrect/don't know
5. C - CP parents know their children's opinion: yes/incorrect/don't know

There are three possible answers to comparisons 4 and 5 (cfr. Kooy, 1969):

person is right about the other's opinion: +
person is wrong about the other's opinion: -
person states he does not know the other's opinion: O
The following comparison gives us the 9 types of mutual knowledge:

6. (P-PC) - (C - CP) mutual knowledge.

Parents	-	Children
+		+
+		O
O		+
+		-
-		+
O		O
-		O
O		-
-		-

7. (P - C) - (C - PC) objective difference of opinion vs subjective difference of opinion of children
8. (P - C) - (P - CP) objective difference of opinion vs subjective difference of opinion of parents
9. (C - PC) - (P - CP) subjective difference of opinion of children vs subjective difference of opinion of parents
10. (P - C) - [(P - PC) - (C - CP)] objective difference of opinion vs types of knowledge.

Example

The example below is taken from one of the 19 items of the list of questions as put before the children. (The list of questions for the parents is the same for this section (= 22 items)). First the children are asked to give their own opinion (a), then they are asked to give their parents' opinion (b). When difference of opinion is apparent the children are also asked whether their parents know that he/she thinks this way (c). As the contents of the relation are not known to us in advance, the items are most generally formulated. The persons concerned shall therefore also be asked to give examples; by means of these facts of experience we shall try to find forms of communication (interaction-strategies). By comparing the answers of children with those of their parents it must become clear whether we can speak of an open, closed, pretence communication or of a communication based on mutual suspicions with regard to this subject. These forms of communication shall be related to the above mentioned 10 comparisons.

Example of one of the questions to children

<table>
<tr><td>a. Do you think your parents do or do not interfere with the upbringing of their grand-children?</td><td>b. What do your parents think of it? Do they think they do or do not interfere with the upbringing of their grandchildren?</td></tr>
<tr><td>In case of difference of opinion
You think that ... (repeat question) and you think your parents do not agree.
Could you give an example of this difference?
Do you notice it? (examples)
Communication
Do you show what you think about this?
Do your parents show you what they think?
Do you ever talk about it?</td><td>c. You think that ... (repeat question)
Do your parents know that you think this?</td></tr>
<tr><td>In case of no difference of opinion
c. You think that ... (repeat question). Do your parents know that you think this?</td><td></td></tr>
</table>

Communication If answer to c. is yes: have you ever talked about it to each other? If answer to 1c. is no/don't know: Do you ever show your feelings on this matter: (does not apply/yes/no/no answer/ answer not known/no opinion/ don't know)	

The respondents

In the foregoing we have always talked of the relation between elderly parents and their children. The reader might have supposed that complete families were included in this research. In fact the planning is much simpler: we selected about 600 persons from the municipality of Nijmegen by means of a sample, 300 men and 300 women, married, and from the 40 - 49 age group. From this group about 100 persons were selected who would finally form the group of respondents. Condition for the selection is: both or at least one of the parents must still be alive and living independently. At the end of the interview with each of these persons (married children of middle age, with at least one of their parents still alive and living independently), we shall ask them if they agree to us having an interview with one of their parents.
So the group of respondents included 100 dyads: every dyad consists of father or mother and one of the married children. We selected approximately a same number of couples: father - son; mother - son; father - daughter; mother - daughter.
The respondents are told beforehand that a different person will interview parents and children.

APPENDIX

Questionnaire-items (as they appear on the questionnaire for the children)

1.a. Do you think there are things that should be changed in the living accomodation of your parents?
b. What do your parents think of it? Do they think there are things that should be changed in their living accomodation?

2.a. Do you think your parents have a lot of or little trouble in spending their free time?
b. What do your parents think of it? etc.

3.a. Do you think your parents have a large or a small financial shortage?
b. What do your parents think of it? etc.

4.a. Do you think your parents spend their money wisely?
b. Do your parents think you spend your money wisely?

5.a. Do politics play an important or an unimportant part in the lives of your parents?
b. Do politics play an important or an unimportant part in your life?

6.a. What political party has your preference?
b. What political party has the preference of your parents?

7.a. Does religion play an important or an unimportant part in the lives of your parents?
b. Does religion play an important or an unimportant part in your life?

8.a. Do you think your parents do or do not interfere with the upbringing of their grandchildren?
b. What do your parents think of it? Do they think they do or do not interfere with the upbringing of their grandchildren?

9.a. Do you think that you bring up your children in a strict/free way?
b. What do your parents think of it? Do they think you bring up your children in a strict/free way?

10.a. Do you think your parents spoil their grandchildren?
b. What do your parents think of it? Do they think they spoil their grandchildren?

11.a. Do you think parents need much or little help from their children?
b. What do your parents think of it?

12.a. Do you think your parents are or are not still able to do a lot themselves?
b. What do your parents think of it?

13.a. If your parents need help, who can help best in your opinion?
- own children
- relatives
- neighbours and acquaintances
- others namely ....

b. What do your parents think of it?

14.a. Do you prefer visiting your parents or do you prefer them visiting you?

b. What do your parents prefer? Do they prefer visiting you or do they prefer you visiting them?

15.a. Do you like visiting your parents?

b. Do your parents like visiting you?

16.a. Do you think your relation with your parents has become better or worse, compared to the past?

b. What do your parents think of it? Do they think their relation with you has become better or worse, compared to the past?

17.a. If you don't agree with your parents, do you show this?

b. If your parents don't agree with you, do they show this?

18.a. What is your opinion on the following statement: The older they become, the more they fall back on their children.

b. What do your parents think of it?

19.a. Do you feel you are well or badly acquainted with what your parents do all day?

b. Do you feel your parents are well or badly acquainted with what you do all day?

20.a. How do you prefer to spend days like Christmas and New Year's Day?
- at home
- at your parents'
- at your own home with your parents
- with friends and acquaintances
- in another way, namely ......
- answer not known
- no opinion, don't know

b. How do your parents prefer to spend days like Christmas and New Year's Day?

ABSTRACT

This paper presents the design of a two sided explorative research on intergenerational relations. To obtain more insight in the qualitative aspects of the intergenerational relation, we want to investigate the subjective engagement of both elderly parents and their adult children to aspects of their mutual relationship. We are interested in the (a)symmetrical character of this relationship with reference to values, norms and opinions and the mutual awareness thereof. The main question is: to what extent and on what points of the relationship have the elderly parents and their children mastered the techniques of role-taking. Therefore, we shall have to propose the question as to what parents and children do and do not know about each other. The types of mutual awareness have to be interpreted in the context of the theories on nuclear family and the position of the elderly in modern society.

REFERENCES

Cage, N.L., L.J. CRONBACH., Conceptual and Methodological Problems in Interpersonal Perception, *Psychological Review*, vol. 62 (1955) 411-422.

Glaser, B.G., A.L. Strauss, Awareness Context and Social Interaction, *American Sociological Review*, vol. 29, 1964, 669-679.

Glaser, B.G., A.L. Strauss, Awareness of Dying.

Goffman, E., *Frame Analysis*, An Essay on the Organization of Experience, Cambridge (Mass.), 1975.

Ichheiser, G., Misunderstandings in Human Relation. A Study in False Social Perception, *American Journal of Sociology*, vol. 55, 1949, 2.

Ichheiser, G., *Appearances and Realities*, Misunderstandings in Human Relations, San Francisco, 1970.

Kooy, G.A., *Het huwelijk in Nederland*, (Spectrum) Utrecht, Antwerpen, 1969.

Mead, G.H., Mind, Self and Society, University Chicago Press, 1934[1], 1970.

Newcomb, M.Th., R.H. Turner, P.E. Converse, *Social Psychology*, The Study of Human Interaction, Holt, Rinehart and Winston, New York, 1965.

Scheff, T.J., Toward a Sociological Model of Consensus, *American Sociological Review*, 32, 1967, 32-46.

Siegrist, J., *Das Consensus-Modell*, Studien zur Interaktionstheorie und zur kognitiven Sozialisation, (F. Enke Verlag), Stuttgart, 1970.

Simmel, G., *Soziologie*, Berlin (Duncker & Humblot), 1890[1], 1968[5].

Stryker, S., Role Taking Accuracy and Adjustment, *Sociometry*, 1957, 286-296.

Tartler, R., *Das Alter in der modernen Gesellschaft*, Stuttgart, 1961.

Weber, M., *Gesammelte Aufsätze zur Wissenschaftslehre*, Tübingen, 1922, (4 Aufl. 1973).

Zijderveld, A.C., *De theorie van het symbolische interactionisme*, Boom, Meppel, 1973.

List of contributors

Monique ASIEL, Dr. Med., Assistant professor Université Libre de Bruxelles (Ecole de Santé Publique, Campus Erasme, route de Lennick 808, 1070 Brussels, Belgium)

Anton BEVERS, Research-Associate in Social Gerontology, University of Nijmegen (Adelbertuslaan 4, Nijmegen, The Netherlands)

Bill BYTHEWAY, Ph.D., Senior Research Fellow, Medical Sociology Research Center (University College, Parkstreet, Swansea, SA 1 6 AZ South Wales)

Alexandru CIUCA, Dr., Scientific Director of the National Institute of Gerontology and Geriatric, Bucarest (Str. Aviator Popa Marin 3, Bucharest, Romania)

Lena DAHL, Psychologist M.A., Research psychologist, Institutet für Gerontologi, Jönköping (Brunnsgatan 30, S-552 55 Jönköping, Sweden)

Margaret DIECK, Head of the Department Deutsches Zentrum für Altersfragen (Rankestrasse 17, D-1000 Berlin 30, West Germany)

Gilbert DOOGHE, Dr., Senior Research Fellow at the Population and Family Study Centre, Brussels. Scientific Supervisor of research projects on Aging (Centrum voor Bevolkings- en Gezinsstudiën, Manhattan Center, Toren H2, Kruisvaartenstraat 3, 1000 Brussels, Belgium)

Jan HELANDER, Ph.D., Psychology, Head of the Gerontology Center, Lund, Sweden (Gerontologiskt Centrum, Karl XI gatan 4, 222 22 Lund, Sweden)

Kees KNIPSCHEER, Teaching and Research Fellow in Family Sociology and Social Gerontology, University of Nijmegen (Adelbertuslaan, 4 Nijmegen, The Netherlands)

Else MELIN, Dr., Head psychologist at Gerontology Center, Lund, (Gerontologiskt Centrum, Karl XI gatan 4, 222 22 Lund, Sweden)

Hennig OLSEN, Economist, Research Associate, Danish National Institute of Social Research (Borgergade 28, 1300 Copenhagen, Denmark)

Paul PAILLAT, Dr. in Law, Head Department of Social Demography, National Institute for Population Studies, France (I.N.E.D., 27, rue du Commandeur, 75675 Paris Cedex 14, France)

Dan SCHLETTWEIN-GSELL, Ph.D. Dr. med., Head of research unit, Stiftung für experimentelle Alternsforschung Felix Platter Spital, 4055 Basel, Switserland (Socinstrasse 32, 4051 Basel, Switserland)

Nan STEVENS, Ph.D., Developmental Psychology/Gerontology, Teaching fellow, Department of Social Gerontology (Erasmuslaan 17, Nijmegen, The Netherlands)

Lars TORNSTAM, Ph.D., Institution of Sociology, Uppsala (Drottning gatan 1A, S-752 20 Uppsala, Sweden)

Andrzej TYMOWSKI, Dr. Econ. Sc., Associate Professor of Social Policy, Chief of Research Department in the Institute of Home Trade and Services (Mianowskiego 15/87, Warschau 02-044, Poland)

L. VANDERLEYDEN, Research associate at the Population and Family Study Centre, Brussels (Centrum voor Bevolkings- en Gezinsstudiën, Manhattan Center, Toren H2, Kruisvaartenstraat 3, 1000 Brussels, Belgium)

Hannah WEIHL, M.A., is Senior Lecturer, Paul Baerwald School of Social Work of the Hebrew University and Director of a sociological research unit at the A.J.D.C., Brookdale Institute (Mendelestr. 14, Jerusalem, Israel)

Halina WORACH-KARDAS, Dr. Assistant-professor (Murarska 9m. 5, 91-465 Lodz, Poland)

List of participants to the Dubrovnik meeting - 1976

1. ASIEL M. (Belgium)
2. BERG S. (Sweden)
3. BEVERFELT E. (Norway)
4. CIUCA A. (Rumania)
5. DAHL L. (Sweden)
6. DIECK M. (W.-Germany)
7. DOOGHE G. (Belgium)
8. HELANDER J. (Sweden)
9. HUET J.A. (France)
10. JACOBSOHN D. (Israel)
11. JALOWIECKE S. (Poland)
12. JYRKILÄ F. (Finland)
13. KNIPSCHEER K. (The Netherlands)
14. MAJCE G. (Austria)
15. MALMBERG B. (Sweden)
16. MELIN E. (Sweden)
17. MEDSTRÖM D. (Sweden)
18. MUNNICHS J.M.A. (The Netherlands)
19. NYGÅRD A.M. (Norway)
20. OLSEN H. (Denmark)
21. ØSTERGÅRD F. (Denmark)
22. PAILLAT P. (France)
23. PAJTAS M-L (Finland)
24. SEVERINSEN C. (Denmark)
25. SHANAS E. (U.S.A.)
26. SMOLIĆ-KRKOVIĆ N. (Yugoslavia)
27. SOLEM P-E (Norway)
28. SVANBORG A. (Sweden)
29. WEIHL A. (Israel)
30. WORACH-KARDAS H. (Poland)

List of participants of the Ystad meeting - 1977

1. ASIEL M. (Belgium)
2. BEVERS A. (The Netherlands)
3. BYTHEWAY W.R. (Great Britain)
4. CIUCA (Rumania)
5. DAHL L. (Sweden)
6. DIECK M. (W. Germany)
7. DOOGHE G. (Belgium)
8. FOLLIN A. (Sweden)
9. HELANDER J. (Sweden)
10. HELLAND H. (Norway)
11. HELLIE HUYCK M. (U.S.A.)
12. JYRKILÄ F. (Finland)
13. KNIPSCHEER K. (The Netherlands)
14. LAASSEN B.K. (Holland)
15. MÅRTENSSON E. (Sweden)
16. Mc CAMISCH C. (Sweden)
17. MELIN E. (Sweden)
18. NATHAN T. (Israel)
19. NYGÅRD A.M. (Norway)
20. OLSEN H. (Denmark)
21. ONKILA A. (Finland)
22. PAILLAT P. (France)
23. PLASCHKE J. (W. Germany)
24. ROTHSTEIN U. (Sweden)
25. SEVERINSEN C. (Denmark)
26. STEVENS N. (The Netherlands)
27. TJÄLLDEN A. (Sweden)
28. TORNSTAM L. (Sweden)
29. TYMOWSKI A. (Poland)
30. VIG G. (Norway)
31. WEIHL H. (Israel)
32. WORACH-KARDAS H. (Poland)
33. ZITOMERSKY J. (Sweden)